PDE Valuation of Interest Rate Derivatives

From Theory to Implementation

First Edition

With 24 illustrations

PETER KOHL-LANDGRAF

First Printing: September, 2007

The content of this book is based on the author's diploma
thesis, submitted to the Faculty of Mathematics and Physics of
the University of Bayreuth, Germany in January 2007.

Prepared with LaTeX and KOMA-Script.
Illustrations were created by using METAPOST.

ISBN: 9-783-83349537-3

Herstellung und Verlag:
Books on Demand GmbH, Norderstedt

Bibliographische Information der Deutschen Nationalbibliothek: Die
Deutsche Nationalbibliothek verzeichnet diese Publikation in der
Deutschen Nationalbibliographie. Detaillierte bibliographische Daten
sind im Internet unter http://dnb.d-nb.de abrufbar.

TO EVA AND JULIUS

Contents

I. Theory and Models   **1**

1. Foundations   **3**

     1.1. Stochastic Processes   3

     1.2. Markov Property   7

     1.3. Brownian Motion   11

     1.4. Ito Calculus   14

     1.5. SDE's and Probability Distributions   20

     1.6. Changing Probability Measures: Girsanov's Theorem . .   23

     1.7. Connection to PDEs: The Feynman-Kac Theorem . . .   26

     1.8. Applications in Finance   31

2. Fixed Income Markets   **37**

     2.1. The Yield Curve   38

     2.2. Interest Rate Securities   42

     2.3. Interest Rate Derivatives   45

     2.4. General Modeling Approach   50

3. Models of the Yield Curve   **59**

     3.1. A Summary of Short Rate Models   60

     3.2. The Heath-Jarrow-Morton Framework   63

     3.3. The Libor Market Model   70

4. Markovian Representations of the Yield Curve   **77**

     4.1. Separable Volatility: The Cheyette Model   78

     4.2. The Analytical Bond Price   85

     4.3. The Valuation PDE   89

     4.4. The Case of Constant Parameters   91

     4.5. Multi-Factor Volatility   95

II. Numerics and Implementation   97

5. Numerical Solution   99
  5.1. Discretization of Differential Opterators 100
  5.2. Finite Difference Schemes 103
  5.3. Consistency, Stability and Convergence 114
  5.4. Alternating Direction Implicit Schemes (ADI) 123
  5.5. Treatment of Boundary Conditions 136
  5.6. Digression: Working on Nonuniform Grids 145

6. Practical Considerations   149
  6.1. Early Exercise Products and Optimal Control Problems 151
  6.2. Local and Stochastic Volatility Specifications 153
  6.3. True Stochastic Volatility 159
  6.4. Calibration to Market Data 161
  6.5. Model Implied Volatility 162

7. Design Issues and C++ Implementation   167
  7.1. Components of the Finite Difference Scheme 168
  7.2. The Valuation Model 176
  7.3. Product Valuation Routines 180

III. Appendix   i

A. Additional Calculations   iii
  A.1. The Black-Scholes Model in a Nutshell   iii
  A.2. Evaluating Expected Values and Black's Formula   iv
  A.3. Expressing a Caplet in Zero-Coupon Bond Puts   v
  A.4. Expressing a Swaption as a Sum of Put Options   vi
  A.5. Product Specific Valuation PDEs vii
  A.6. Finite Differences: Further Discretizations viii

B. Probability Essentials   ix

List of Figures

1.1. Stochastic Time Evolution of Brownian Motion 12

2.1. A Forward Term-Rate represented $F(t, T_1, T_2)$ as a Zero-Coupon Bond Portfolio 41

3.1. Evolution of the Yield Curve driven by Stochastic Forward Rates 64
3.2. Drift-Evolution in the Libor Market Model 75

5.1. Grid points (black) involved in the Explicit, Implicit and Weighted Scheme to calculate grid points in red. 108
5.2. Correct Resolution of the Flow along the Characteristics with $a > 0$ and $a < 0$ by the Upwind Scheme 109
5.3. Second Space Dimension of the Zero-Coupon Bond $P(t, T; x, y)$ (4.10) with Upwind and Centered Discretization versus Analytical Solution. 112
5.4. Alternating Implicitness of the PR-Scheme 125
5.5. Splitting Procedure in the 2D CS-Scheme with $0 < \theta < 1$ 128
5.6. Numerical Error of Linear Extrapolation at the boundaries might introduce an error (grey area) in view of the evaluation of $V(0, x) = \mathbb{E}[\Phi(X_T)|X_0 = x]$ 140
5.7. Effect of Wrong Boundaries: First Figure with $[-0.2, 0.2]$ and Second Figure with $[-0.88, 0.88]$ interval for the first state variable 144
5.8. The Grid Transformation (left) and a dense grid (right) 147

6.1. Valuation PDE (4.13): Model Implied Volatility is flat for $m = 0$ and skewed for $m = 1$ 164
6.2. Extended Model (6.5): The Model's Implied Volatility Smile for $m = 0$ and additionally skewed for $m = 1$ . . 165

7.1. Class Diagram for the Finite Difference Scheme 168

7.2. UML Diagram for the ADI Scheme 169
7.3. The Grid Class 171
7.4. Classes for Boundary Specification and Handling 172
7.5. UML Diagram for the Cheyette Model 176
7.6. The Product Class for the Zero-Coupon Bond 180
7.7. The Product Class for the Caplet 181
7.8. The Product Class for the Interest Rate Swap 182
7.9. The Product Class for the European Swaption 183
7.10. The Product Class for the Bermudan Swaption 184
7.11. C++ Design - Core Classes at a Glance 185

Preface

Purpose of the book

Financial Quantities are valued in a specific model environment under the use of two main assumptions. The first one is of economical nature and postulates that it is not possible to generate riskless profit when trading in financial securities or financial derivatives as contingent claims. The second concerns the mathematical modeling environment and assumes the quantity to behave as a stochastic process over time to meet the aspect of future uncertainty.

These two combined assumptions were firstly successfully applied to value an option on an equity stock by F. Black and M. Scholes (1973). In the early nineties (1992) this valuation approach was also standardized for the interest rate world by D. Heath, R. Jarrow and A. Morton (HJM), who determined the risk-neutral dynamics of the yield curve.

The HJM Model has shown to be the most general modeling environment in the interest-rate world, as it incorporates many other so-called short-rate interest rate models which came historically first with the Vasicek Model in 1977. However it suffers from two main disadvantages: This general setting is difficult to apply in market practice and the structure of the underlying high-dimensional stochastic process involves an increased computational complexity.

The first disadvantage was overcome by the introduction of the Libor Market Model (1997) which combined the general risk-neutral yield-curve model with market standards. Through that the Libor Market Model and its several extensions can be seen as the main reference in interest rate modeling until today.

As not having resolved the second disadvantage and due to the ever increasing complexity of interest rate products the Market Model Approach starts to reach its limits from the computational side.

This book is mainly concerned with a class of interest rate models which try to overcome the second mentioned disadvantage by breaking down the general HJM Model to a subclass with desirable numerical features. These model can be derived directly out of the HJM Model by a specification of the process structure. A SDE System emerges which offers nice features from a numerical perspective as it breaks down the high-dimensional HJM dynamics into a low-dimensional Markovian structure. Furthermore through the process' dependence on the particular current state it enables the modeler to do the valuation backwards in time by a solution of a certain Partial Differential Equation (PDE).

Although this approach was firstly developed in 1992 by O. Cheyette in [1], it has been fallen into oblivion due to the overwhelming success of the Libor Market Model. However it has been recovered recently in a more general setting by going *Back to the Future* in [2].

The objective of this book is threefold:

⋄ To illuminate in a compact way the connection between stochastic processes and partial differential equations

⋄ To embed the here analyzed Markovian model class in the entire framework of interest rate models

⋄ To present and implement a robust numerical solver, which enables an efficient computational treatment of product valuation in this specific model setup.

How this book is organized

The first chapter is concerned with a review of results from stochastic calculus with a specification on Brownian Motion and Ito Calculus. The main results are the Girsanov Theorem, which theoretically enables a risk-neutral valuation under a measure transformed stochastic process and the Feynman-Kac Theorem, which builds the basis for the valuation of future financial payoffs to be done by the use of partial differential equations. We sum up this chapter with a general review on applications in finance and with an emphasis on risk-neutral valuation.

In the second chapter we give a general overview of the fixed-income market by defining several interest rate quantities including the yield curve. We introduce common interest rate securities and plain as well as exotic interest rate derivatives. Furthermore we point to the general modeling approach under stochastic interest rates and the market standard, represented by Black's Model, which always has to be taken into consideration when investigating a new model setup.

All general approaches to model interest rate quantities are discussed in Chapter 3 with a review of short-rate models and a general derivation and discussion of the one-factor HJM yield curve dynamics. We also provide a derivation of the Libor Market Model directly out of the general HJM setting.

In Chapter 4 we derive and discuss the Markovian Model setup through a specification of the volatility structure in the HJM Model. We derive the analytical bond price formula and the valuation PDE for general interest rate products under the use of the Feynman-Kac Theorem. With certain parameter specifications we also show, how this model setup is related to other short-rate models.

In view of the numerical approach to solve the resulting two-dimensional valuation PDE the finite difference approach including consistency and stability analysis is reviewed in Chapter 5 in combination with an operator splitting algorithm known as the ADI Method. This is followed up by a discussion on how to approximate the PDE's boundary behavior, as there is now explicit boundary condition available. The chapter is ended up with a digression on finite differences on non-equidistant grids.

The sixth chapter reviews practical advantages of the theoretical model setup from Chapter 4. Due to the dependence only on the particular current system state we are able to apply Bellmann's Optimality Principle and apply the Backward Optimization Algorithm for the valuation of products equipped with an early exercise feature. We also give an overview of how the model is used in practice with several parameter specifications. An extended three-dimensional model setup - and the corresponding three-dimensional valuation PDE - under the use of stochastic volatility is discussed. Furthermore it is stressed, why in the context of stochastic volatility the model is preferable to other short-rate models.

In view of calibration purposes we analyze the developed valuation PDE's in view of their general quality and flexibility to match diverse market phenomena - i.e. implied volatility smiles and skews.

The last chapter is concerned with the C++ implementation side of the finite difference scheme from Chapter 5 in combination with the valuation model from Chapter 4. After having explained this back-end component we show how this can be used to value several interest rate products.

In the appendices we provide additional calculations and basic results from probability theory.

Part I.

Theory and Models

1. Foundations

To work within a proper framework we firstly give a summary on some results from stochastic calculus with a short outlook on their applications in finance. We make an early specification to Brownian Motion as the main driver of uncertainty in financial theory.

The developed concepts lead to two main theorems, which are sufficient for the needs in this thesis concerning valuation of financial contingent claims. In particular the Feynman-Kac Theorem in Section 1.7 establishes the theoretical basis for a deterministic PDE valuation of financial quantities when modeled as Markovian processes. For basic results and definitions from stochastic theory, which are not explained in the current context, the reader is referred to Appendix B.

1.1. Stochastic Processes

Unless otherwise stated we consider the following probability space $(\Omega, \mathcal{F}, P)$[1], where each $\omega \in \Omega$ will be referred to as a *scenario*.

Definition 1.1. *A n-dimensional Stochastic Process X is a family of random variables $\{X_t\}_{t \in [0,T^*]}$ parameterized through a time-parameter $t \in [0, T^*]$ in the sense that:*

$$X(\cdot) : \Omega \to \mathcal{P}$$
$$X(\cdot, \cdot) : [0, T^*] \times \Omega \to \mathcal{S},$$
$$(t, \omega) \mapsto X(t, \omega) \quad where \quad X_t(\omega) := X(t, \omega) \in S$$

$\mathcal{P}$ is the Path Space *of all possible paths $X(\omega)$, $\mathcal{S}$ is the* State Space *and $[0, T^*]$ with $T^* \in (0, \infty)$ denotes for our purposes a finite continuous or discrete time interval.*

Remark 1.1. *The state space may be time-dependent and restricted. Unless otherwise stated we set $\mathcal{S}$ to be $\mathbb{R}^n$.*

---

[1]Probability Space: See Appendix B

For each scenario $\omega \in \Omega$, $X(\omega) \in \mathcal{P}$ is called the *Sample Path* as a function $t \mapsto X_t$ and for each fixed time-point $t \in [0, T^*]$ we term the resulting random variable $X_t(\omega) \in \mathcal{S}$ *event* or *outcome* based on scenario ω. In the continuous-time setting we specify the path space $\mathcal{P}$ to be the function space of continuous functions $\mathcal{C}([0, T^*])$.

Information structure

Since the random variables become values, when time is proceeding, the information about the stochastic process changes over time. If more information about the process gets revealed, the probability of occurrence on other events and therefore the probability measure has to change. To circumvent working with time-changing probability measures we now introduce a concept, which describes the information structure or history generated by a stochastic process up to a certain point.

Definition 1.2. *A* Filtration *on* $(\Omega, \mathcal{F})$ *is a family of increasing σ-algebras* $\mathcal{F} = \{\mathcal{F}_t\}_{t \geq 0}$ *with* $\mathcal{F}_t \subseteq \mathcal{F}$ $\forall t \in [0, T^*]$ *in the sense that:*

$$s \leq t \Rightarrow \mathcal{F}_s \subseteq \mathcal{F}_t \quad \forall s, t \in [0, T^*]$$

A probability space equipped with such a family is referred to as filtered probability space *and is denoted as* $(\Omega, \mathcal{F}, P; \{\mathcal{F}_t\})$.

As random variables induce σ-algebras, one obvious choice of a filtration would certainly be the σ-algebra generated by a stochastic process up to time t itself:

$$\mathcal{F}_t^X = \sigma(X_s | 0 \leq s \leq t)$$

The essential advantage of a filtration is that we can let the probability measure P stay fixed and instead of taking an expectation with time-changing measures, we now take the *P-Conditional Expectation*[2] with respect to that increasing σ-Algebra: $Y_t := \mathbb{E}^P[X_T | \mathcal{F}_t]$. This conditional expectation parameterized by the filtration introduces itself a new stochastic process Y_t.

Given such an information structure we can make expectations on other

---

[2]Conditional Expectation: See Appendix B

outcomes, so we further define, what kind of events are measurable with respect to a existing information structure:

Definition 1.3. *A stochastic process X is called* adapted *to a filtration $\mathcal{F}_t$, if X_t is measurable w.r.t. $\mathcal{F}_t{}^3$.*

This basically says that knowing the outcomes X_s up to time t is sufficient to determine the probabilities of possible outcomes X_t. If the referred filtration is unique, we simply call a process adapted.

Martingales

To be more specific we now introduce *two* classes of stochastic processes, where we show later on that Brownian Motion is an instance of those. Processes of the first class have the meaning that the best guess we can make on future values given all the information up to a time t, is the presently known one.

Definition 1.4. *A $\mathcal{F}_t$-adapted stochastic process $M = \{M_t\}_{t \in [0,T^*]}$ is said to be a* Martingale, *if $\forall t \in [0, T^*]$:*

$$[\text{i}] \quad \mathbb{E}^P\big[|M_t|\big] < \infty \tag{1.1}$$

$$[\text{ii}] \quad \mathbb{E}^P\big[M_T|\mathcal{F}_t\big] = M_t \tag{1.2}$$

The second condition can also be rewritten as: $\mathbb{E}^P\big[M_T - M_t|\mathcal{F}_t\big] = 0$.

The following Proposition provides an easy way to construct martingales as a conditional expectation of given random variables.

Proposition 1.1. *Let there be an $\mathcal{F}_t$-adapted random variable $H : \Omega \mapsto \mathbb{R}^n$ with*

$$\mathbb{E}^P[|H|] < \infty$$

Then the process $M = \{M_t\}_{t \in [0,T^*]}$ defined by $M_t := \mathbb{E}[H|\mathcal{F}_t]$ is a martingale.

---

[3]Measurable: See Appendix B

Proof. This follows from the *Tower Property*[4] of conditional expectation:.

For some $t < T \in [0, T^*]$ we have:

$$\mathbb{E}[M_T | \mathcal{F}_t] = \mathbb{E}[\mathbb{E}[H | \mathcal{F}_T] | \mathcal{F}_t] = \mathbb{E}[H | \mathcal{F}_t] = M_t$$

$\square$

---

[4]Tower Property: See Appendix B.15

1.2. Markov Property

The next here discussed important class of stochastic processes has the interpretation of being in some sense memoryless as those preserve all past information at any particular current state. So given a Markovian Process, if we'd like to make a guess at a particular time about the future value with respect to all past information, the knowledge of the process' outcome at that current time was sufficient to do so.

We give here a short review of this property and collect some important results. We also emphasize in this context the basic connection between such processes and partial differential equations. Those describe the forward dynamics of the respective underlying probability densities and backward dynamics of conditional expectations.

Firstly, we define the Markov Property in terms of conditional expectation.

Definition 1.5. *A stochastic process X is said to be a* Markov Process *if it satisfies the* Markov Property*:*

$$\mathbb{E}[X_\tau | \mathcal{F}_t] = \mathbb{E}[X_\tau | \sigma(X_t)] \quad \forall \tau \in [t, T^*] \tag{1.3}$$

Here $\sigma(X_t)$ denotes the σ-Algebra which is generated by the random variable X_t - respectively the process at time t. Such a process on a discrete time scale $\{t_0, \ldots, t_n\} \subset [0, T^*]$ is referred to as a Markov Chain.

The Markov Property defines so-called Transition Probabilities*:*
If we define outcomes $x_0, \ldots, x_{i-1}$ and $x := x_i$ as elements of a state space $\mathcal{S}$ we then have for conditional probabilities at times t_i, t_{i+1} ($i = 0, \ldots n - 1$):

$$P\big(X_{t_{i+1}} = y \big| X_\tau = x_\tau \ \forall t_0 \leq \tau \leq t_i\big) = P\big(X_{t_{i+1}} = y \big| X_{t_i} = x\big) := p_{xy}^{t_i}$$

Furthermore a Markov Chain is called Stationary *or* Homogeneous[5]*, if $p_{xy}^{t_i} \equiv p_{xy} \ \forall i$.*

---

[5]A more precise definition of Stationarity is given in Definition 1.9

There are two main equations related to Markov Chains which we firstly introduce here on a discrete state space $\mathcal{S}' \subset \mathbb{R}$ and discrete time-scale $\{t_0, \ldots, t_n\}$.

Definition 1.6. Forward Equations *from a time t_i to time t_{i+1} describe the dynamics of transition probabilities:*

$$
P(X_{t_{i+1}} = y) = \sum_{x \in S'} P(X_{t_{i+1}} = y | X_{t_i} = x) P(X_{t_i} = x)
$$

$$
= \sum_{x \in S'} p_{xy}^{t_i} P(X_{t_i} = x) \tag{1.4}
$$

Backward Equations *from a time t_{i+1} to time t_i describe expected values of functions $\Phi(.)$ depending on the final outcome X_T of a Markov process at a future time point $T \in \{t_{i+2}, \ldots, t_n\}$:*

$$
\mathbb{E}[\Phi(X_T) | X_{t_i} = x]
$$

$$
= \sum_{y \in S'} \mathbb{E}[\Phi(X_T) | X_{t_i} = x \cap X_{t_{i+1}} = y] P(X_{t_{i+1}} = y | X_{t_i} = x)
$$

$$
= \sum_{y \in S'} \mathbb{E}[\Phi(X_T) | X_{t_{i+1}} = y] p_{xy}^{t_i} \tag{1.5}
$$

with Final-Time Condition $\mathbb{E}[\Phi(X_T) | X_T = x] = \Phi(x)$.

Remark 1.2. *One important specialization of $\Phi(.)$ in (1.5) leads in the continuous time and space limit to the Feynman-Kac PDE (1.34) discussed in Section 1.7.*

Trees as Discretizations of Markov Processes

In the time-discrete version with a countable state space the set of all possible paths can be modeled by a graph (G, E), where each node $v \in G$ corresponds to an element in the state space $\mathcal{S}$. On each edge $e \in E$ we move to the next time point and take a new value in $v \in G$. This tree is called *Decision Tree*. Each sample path starts from the unique root node - if no initial probability distribution is assumed - and ends at one of the multiple end nodes at the end of the time interval.

Since different depths in trees now correspond to different time points, we usually say to move *forward or backward* in the tree.
From a given node at a certain time point t, there are n edges to move forward with time-dependent probabilities p_i^t $(i = 1 \ldots n)$ and $\sum_{i=1}^{n} p_i^t = 1$.

For $n = 2, 3$ this tree is called *Binomial Tree* and *Trinomial Tree* respectively. In the time-homogenous case all moves are identically distributed, so in the continuous time limit this tree will weakly converge in the sense of the Central Limit Theorem (B.11) and is therefore a consistent approximation of the time-continuous stochastic process.

In the case of time-homogeneous Markov chains there are several ways to get up to a certain node, as the Markovian process moves forward in time by neglecting the past.
As all past process information is contained in the value of the actual node, only that value is important for the next move, which is termed to be a *Path Independence*. Due to that fact the decision graph assumes a cyclical structure and is denoted as a *Lattice*.

Important for numerical considerations, in a modeling context it will be always useful to prefer Markovian stochastic processes compared with general ones, since the number of nodes at each time steps grows at a linear asymptotic rate $O(t)$ compared to the exponential growth rate n^t in the general case.

Diffusion Processes

Passing to the continuous space limit of those Markov Chains leads to the notion of *Diffusion Processes*.

Definition 1.7. *Transition probabilities of* Diffusion Processes *are defined via their* Transition Densities[6] $p(\tau, x; t, y)$. *Those determine the probability of arriving in a set B at time $\tau \in (t, T^*]$ (i.e. $X_\tau \in B$) seen from a time $t \in [0, T^*]$ at state $X_t = x$.*

$$P(X_\tau \in B | X_t = x) := \int_B p(\tau, y; t, x) dy \quad with \quad B \subset \mathbb{R}^n \quad (1.6)$$

These time-dependent densities satisfy certain parabolic Partial Differential Equations [PDE's]. We discuss this feature in the special case of the now introduced process family: Brownian Motion.

---

[6]Probability Density: See Appendix B.

1.3. Brownian Motion

As announced in the introduction of the current chapter we now specify
our previous results to a diffusion process with Gaussian distributed[7]
increments. Since not having introduced a continuous time differential we firstly give here a time-discrete definition of its basic properties
following [3, p.29].

Definition 1.8. *A stochastic process* $\{W_t\}_{t \in T}$ *is called P-Brownian
Motion or* Wiener Process, *if it has the following properties*

1. $P(\omega \in \Omega | W_0(\omega) = 0) = 1$ *i.e.* $W_0 = 0$ *w.p.1*

2. *For any finite set of times* $0 \leq t_1 < t_2 < ... < t_n \leq T$
   the discrete time increments $\Delta W_{t_i} := W_{t_i} - W_{t_{i-1}}$ *are independent*

3. *For any* $0 \leq t_i \leq t_j \leq T$: $\Delta W_{t_j} \sim \mathcal{N}(0, t_j - t_i)$

4. $P(\omega \in \Omega | W(\omega)$ *is continuous* $) = 1$

Remark 1.3. *Brownian Motion is a martingale due to the third property. The time independence of its increments makes it a Markovian
process. The time-continuous version of the increments leads to the
notion of the* Ito Integral *discussed in the next section.*

The second property is sufficient for its Time Stationarity:

Definition 1.9. *A stochastic process is said to be* Stationary *or* Homogeneous, *if its joint distribution densities are independent of time shifts.
In particular* X_t *and* $X_{t+\Delta t}$ *have the same density regardless the choice
of the point in time* t.

We now come back to the previous section: Since Brownian Motion is
normally distributed, its transition density, which determines the probability of arriving at space state y at a future time point τ seen from
time point t at state space x, is given in the sense of Definition 1.7 with
$\tau \in (t, T^*]$ by:

$$p^{BM} := p(\tau, y; t, x) = \frac{1}{\sqrt{2\pi[\tau - t]}} \exp\left[-\frac{[x - y]^2}{2[\tau - t]} \right] \tag{1.7}$$

---

[7]for Gaussian/Normally distributed Random Variables and properties: see Appendix B

The time-continuous forward dynamics of this density is given is by direct calculation. The following equation can be seen as an analogue to the previously introduced time and space discrete forward equation (1.4) for transition probabilities.

$$\frac{\partial p^{BM}}{\partial \tau} - \frac{1}{2}\frac{\partial^2 p^{BM}}{\partial y^2} = 0 \tag{1.8}$$

This partial differential equation (PDE) is well known as the *Heat Equation* or *Diffusion Equation*. By this basic connection processes equipped with Brownian Motion increments become connected to partial differential equations equipped with diffusion part.

For backward equations (1.5) - respectively the backward dynamics of conditional expectations - there is also an equivalent equation for its time-continuous dynamics, which we will further discuss later on in the special case of the Feynman-Kac Formula.

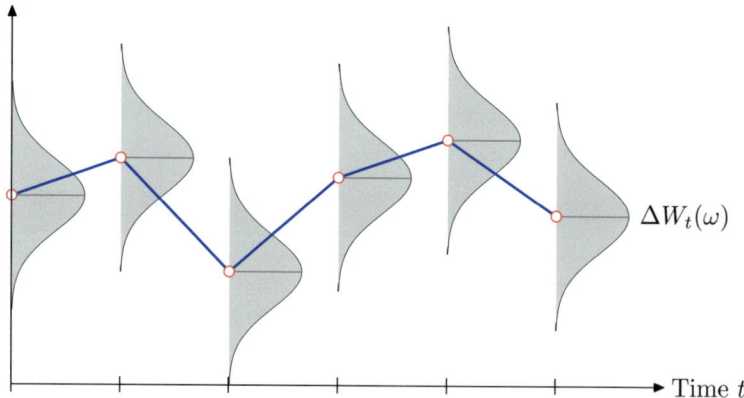

Figure 1.1.: Stochastic Time Evolution of Brownian Motion

Remark 1.4. *A stochastic process X_t defined by:*

$$X_t := \mu_t t + W_t \tag{1.9}$$

with a stochastic process μ_t is referred to as a Brownian Motion with drift and drifless if $\mu \equiv 0$.

Brownian Motion in Multiple Dimensions

Starting with a n-dimensional vector of Brownian Motions

$$W_t = (W_t^1, \ldots, W_t^n) \text{ on } [0, t]$$

which are pairwise independent, we can establish a correlation between those random variables in the sense of a linear combination regarding a chosen vector space, represented by a matrix $A \in \mathbb{R}^{n \times n}$ by setting:

$$\tilde{W}_t = AW_t$$

This yields for the Covariance Matrix[8] - usually also called *Auto Covariance*:

$$[C]_{i,j} = Cov[\tilde{W}_t^i, \tilde{W}_t^j] = < A_i | A^j > t$$

with Correlation Matrix:

$$[\rho]_{ij} = \frac{< A_i | A^j >}{< A_i | A^i >} \quad \forall i, j = 1, \ldots, n$$

Here $< A_i | A^j >$ denotes a scalar-product of the ith row and the jth column of a matrix A.

---

[8]Covariance and Correlation: See Appendix A

1.4. Ito Calculus

To mathematically model dynamics under stochastic perturbation Ordinary Differential Equations [ODE's] can be equipped with continuous stochastic processes, which leads to the notion of *Stochastic Differential Equations* [SDE's]. Usually this perturbation is modeled with increments of a Brownian Motion ΔW_t from above. Start with some discretized dynamics in form of a difference equation:

$$X_{t_k}(\omega) - X_{t_{k-1}}(\omega) = f(t_k, \omega)\Delta_{t_k} + g(t_k, \omega)\Delta W_{t_k}(\omega)$$

where $f(t, \omega)$ and $g(t, \omega)$ are some stochastic processes. Building the difference quotient and taking time limit provides a major difficulty, since there is up to now only a discrete definition of Brownian Motion increments and, even worse, these are not differentiable (w.p.1). and therefore $\lim_{\Delta t \to 0} \frac{\Delta W_t}{\Delta t}$ would not be well-defined. Therefore in contrast to ordinary calculus, the stochastic equivalent introduces first the notion of a *Stochastic Integral* and afterwards implicitly constructs the stochastic differential.

We concentrate here on the construction of the *Ito Integral*. With this notion the time limit of the summed-up difference equation becomes a stochastic process, which is referred to as the *Ito Process*:

$$X_T(\omega) = X_0(\omega) + \int_0^T f_\tau(\omega)d\tau + \underbrace{\int_0^T g_\tau(\omega)dW_\tau(\omega)}_{\text{Ito Integral } I[g]} \quad \text{with } T \in [0, T^*]$$

Brownian Motion does not have the property of *Bounded Variation*, which makes it difficult to define the limit in the sense of the Riemann integral. It is therefore constructed in a squared sense using the fact that Brownian Motion does have the property of *Finite Quadratic Variation* (see [4, p.56]).

The start point is to define the integral on a discretized partition on a finite-sub interval $[T_1, T_2] \subset [0, T^*]$:

$$T_1 = t_0^n \leq t_1^n \leq t_2^n \leq \ldots t_n^n = T_2 \text{ with } \Delta_i^n := t_i^n - t_{i-1}^n$$

Corresponding to the Regular Functions which are used to construct the Riemann Integral Version, we introduce *Step Processes* $\bar{g}_t(\omega)$ in the sense that they denote one random variable for each time interval in the following sense:

$$\bar{g}_t^n(\omega) \equiv g(t_j^n, \omega) \quad \forall t \in [t_i^n, t_{i+1}^n) \quad (i = 0, \ldots, n-1)$$

With those piecewise constant random variables we get a first discrete formulation of the Ito Integral:

$$I_{T_1,T_2}^n[g](\omega) := \sum_{i=0}^{n-1} \bar{g}_{t_i}^n(\omega) \Delta W_{t_{i+1}^n}(\omega)$$

As we have chosen the most left-end points and because the above sum is a random variable itself taking the conditional expectation yields in combination with the distributional property of Brownian Motion in Definition 1.8[9]:

$$\mathbb{E}\left[I_{T_1,T_2}^n[g]^2 | \mathcal{F}_{T_1}\right] = \sum_{i=0}^{n-1} \mathbb{E}[\bar{g}_i^n(.)^2 | \mathcal{F}_{T_1}] \Delta_{i+1}^n \tag{1.10}$$

Definition 1.10. *Assume that the process g_t is out of a class of stochastic processes, which can be well approximated by the above step processes $\bar{g}_t(\omega)$[10]. On $[T_1, T_2] \subseteq [0, T^*]$ the Ito Integral $I_{T_1,T_2}[g]$ is defined as:*

$$\lim_{n \to \infty} \mathbb{E}\left[\left|I_{T_1,T_2}[g] - I_{T_1,T_2}^n[g]\right|^2\right] = 0 \tag{1.11}$$

We then write for the Ito Integral:

$$\int_{T_1}^{T_2} g_\tau dW_\tau := I_{T_1,T_2}[g] \tag{1.12}$$

A family of Ito Integrals $\{I_{t,u}[g]\}_{u \in [t,T^*]}$ defines itself a stochastic process.

---

[9]Since each summand consists of two independent random variables and Brownian Motion does have zero mean, the formula: $\mathbb{E}[\sum X_i^2]) = \sum \mathbb{E}[X_i^2]$ can be applied.
[10]For further details of this process class see [4, p.25].

The above introduced integral version basically differs in two ways from an ordinary Riemann Integral. Firstly, it does make a difference what base points one chooses in the stepwise approximation to arrive at formula (1.10) and secondly, the Ito Integral is not constructed pathwise as the limit for each scenario ω but as the limit in the above *Mean-Square* sense.

Concluding our discussion on Ito Integrals we emphasize two important features.

Remark 1.5 (Martingale Property). *The parameterized Ito Integral* $\{I_{t,u}[g]\}_u$ *is itself a martingale[11].*

$$\mathbb{E}\Big[I_{t,u}[g]\big|\mathcal{F}_t\Big] = \mathbb{E}\Big[\int_t^u g(\tau,.)dW_\tau(.)\big|\mathcal{F}_t\Big] = 0 \quad \forall u \in (t,T^*] \quad (1.13)$$

Remark 1.6 (Ito Isometry). *Formula (1.10) is also known as the Ito Isometry[12]:*

$$\mathbb{E}\big[I_{t,T}^2|\mathcal{F}_t\big] = \mathbb{E}\Big[\big[\int_t^T g(t,.)dW_t\big]^2\big|\mathcal{F}_t\Big] \overset{1.10}{=} \mathbb{E}\Big[\int_t^T g_t^2(.)dt|\mathcal{F}_t\Big]$$
$$(1.14)$$

Having the Integral now well-defined leads to the notion of Ito Processes.

Definition 1.11. *A* n-dimensional *Ito Process is a stochastic process of the form:*

$$X_t(\omega) = X_0(\omega) + \int_0^t a(s,\omega)ds + \int_0^t B(s,\omega)dW_s(\omega) \quad \forall t \in [0,T^*]$$
$$(1.15)$$

with a m-dimensional *Brownian Motion* $W_t \in \mathbb{R}^m$ *and with* $a_t(.) \in \mathbb{R}^n$ *and* $B_t(.) \in \mathbb{R}^{n \times m}$ *adapted processes, which satisfy boundedness conditions for* $i = 1, \ldots, n$:

$$P\Big(\int_0^T |a_t^i|ds < \infty\Big) = 1 \ \text{and} \ P\Big(\int_0^T |(B_t^T B_t)_i|^2 ds < \infty\Big) = 1$$

---

[11]For a proof of a slightly different version see [4, p.32].
[12]A detailed proof of this result can be found in [4, p.26].

Following [4, p.44] a shorter Differential Form *is usually given in the sense that if X_t follows (1.15), then we write:*

$$dX_t = a_t dt + B_t dW_t \qquad (1.16)$$

With the following *Ito Formula*, which is referred to as the *stochastic equivalent* of the chain rule from ordinary calculus, it is possible to analyze dynamics of functions depending on the outcome of Ito Processes. That becomes essentially important when solving SDE's introduced in the next section.

Theorem 1.1 (Ito's Lemma). *Let X_t be an Ito Process of the form (1.16), and $v(t,x) : [0,T^*] \times \mathbb{R}^n \mapsto \mathbb{R}^m$ twice continuously differentiable. Then the process:*

$$Y_t := v(t, X_t)$$

is again an Ito Process and is given in differential form with $k = 1, \ldots, m$ as:

$$dY_t^k = \frac{\partial v_k}{\partial t}(t, X_t) + \sum_{i=1}^{n} \frac{\partial v_k}{\partial x_i}(t, X_t) dX_i$$

$$+ \underbrace{\frac{1}{2} \sum_{i=1}^{n} \sum_{j=1}^{n} <B_i|B^j> \frac{\partial^2 v_k}{\partial x_i \partial x_j}(t, X_t) dt}_{Ito-Calculus} \qquad (1.17)$$

A detailed proof can be found in [4, p.46] and is based on a Taylor-Expansion of dY_t. Neglecting terms of higher order than $O(\Delta t)$ and using the Ito Isometry (1.14) yields in contrary to ordinary calculus to the additional third term with second order partial derivatives.

Remark 1.7. *Following [3, pp.126] we introduce a dimension independent notation:*

$$dY_t = v_t(t, X_t)dt + v_x(t, X_t)dX_t + \frac{1}{2}v_{xx}(t, X_t)dX_t dX_t \qquad (1.18)$$

An extension of the above stochastic chain rule for functions depending on two processes is given by the following Product Rule - for a more general form see [5, p.127].

Theorem 1.2 (Ito Product Rule). *Consider two Ito Processes X_t^i with $f^i(t, \omega)$ and $g^i(t, \omega)$ $(i = 1, 2)$ as in (1.15).*
Then the product process $Y_t := v(X_t^1, X_t^2) := X_t^1 X_t^2$ is again an Ito Process where its differential form is given by the following Product Rule*:*

$$dY_t = X_t^1 dX_t^2 + X_t^2 dX_t^1 + dX_t^1 dX_t^2 \qquad (1.19)$$

Ito Processes as solutions of SDE's

With the notion of an Ito Process (1.15) a well-defined framework for treating stochastic differential equations [SDE's] has been developed, where the Ito Formula (1.1) provides a concept to solve them.

As we have considered Ito Processes with general stochastic coefficients a_t and B_t as in (1.15), we proceed considering coefficients depending only on time and outcome of the process X_t at time t itself.

Definition 1.12 (Markovian SDE & Ito Diffusion). *Consider a n-dimensional Stochastic Differential Equation (SDE) whose solution corresponds to an Ito Process $\{X_t\}_t$ of the form (1.15).*

$$dX_t = a(t, X_t)dt + \underbrace{\Sigma(t, X_t)C}_{:=\sigma(X_t, t)} dW_t \qquad (1.20)$$

The evolution of this kind of SDE's depends only on the current state X_t and therefore satisfies the Markov Property (1.3). The resulting Ito Processes as solutions of these equations are known as Ito Diffusions. *Here CdW_t denotes a n-dimensional Brownian Motion $W_t \in \mathbb{R}^n$ of correlated scalar Brownian Motions W_t^i with $C \in \mathbb{R}^{n \times n}$ in the sense that:*

$$[dW_t]^T[C^T C][dW_t] = [C^T C]dt$$

and with the notions of the Drift-Vector $a \in \mathbb{R}^n$ *and the* Diffusion *Matrix $\sigma \in \mathbb{R}^{n \times n}$. This matrix is also called* Volatility Matrix *in the financial context.*

Remark 1.8. *SDE's (respectively Processes X_t as corresponding solutions) whose drift and diffusion coefficients depend on states X_τ earlier than the current state X_t are termed* Path Dependent. *These kinds of processes do* not *satisfy the Markov Property (1.3).*

Remark 1.9. *With its drift term being zero $a \equiv 0$ the SDE is said to be driftless. The associated process is a martingale due to property (1.13).*

1.5. SDE's and Probability Distributions

Analytically solvable SDE's reveal direct information about their stochastic distributional properties. In contrary to ODE's, where the corresponding solution would be a deterministic function, the solution of a SDE is simply a stochastic process, whose probability distribution - due to the Brownian Motion Term - is usually closely related to the normal distribution. In this section we give two examples of SDE's for which we can analytically determine the distributions of their corresponding solutions and which furthermore play a crucial role in the context of modeling financial quantities such as equity stocks or interest rates.

Proposition 1.2. *The Markovian process as the solution of the SDE given through:*

$$\boxed{dX_t = \mu X_t dt + \sigma X_t dW_t} \tag{1.21}$$

is termed Geometric Brownian Motion. *The process is lognormal distributed.*
For given X_t the conditional distribution for X_T with $T \geq t$ is given by:

$$\log X_T \sim \mathcal{N}\left(\log X_t + (\mu - \frac{1}{2}\sigma^2)(T - t), \sigma^2(T - t)\right) \tag{1.22}$$

The transition density which determines this distribution corresponds to a slightly modified version of (1.7).

Proof. Applying Ito's Lemma to $d[f(X_t)]$ with $f(x) := \log x$ yields:

$$d \log X_t = \left(\mu - \frac{1}{2}\sigma^2\right)dt + \sigma dW_t$$

Integration and rearranging yields the solution of the SDE:

$$X_T = X_t \exp\left((\mu - \frac{1}{2}\sigma^2)(T - t) + \int_t^T \sigma dW_\tau\right) \tag{1.23}$$

The mean of the above distribution (1.22) is therefore obvious where the variance follows straightforward from (1.14).

$\square$

Proposition 1.3. *The Markovian process as the solution of the SDE given through:*

$$dX_t = (b - aX_t)dt + \sigma dW_t \tag{1.24}$$

is termed Ornstein-Uhlenbeck Process. *It is normally distributed with conditional mean:*

$$\mathbb{E}[X_T|\mathcal{F}_t] = X_t e^{-a(T-t)} + \frac{b}{a}\left(1 - e^{-a(T-t)}\right) \tag{1.25}$$

and conditional variance:

$$\mathrm{Var}[X_T|\mathcal{F}_t] = \frac{\sigma^2}{2a}\left(1 - e^{-2a(T-t)}\right) \tag{1.26}$$

Proof. Solving this SDE explicitly is done by firstly observing that:

$$d\left[e^{at}X_t\right] = e^{at}dX_t + ae^{at}X_t dt = be^{at}dt + e^{at}\sigma dW_t$$

Time-integration from t to T yields:

$$e^{aT}X_T = e^{at}X_t + \int_t^T be^{a\tau}d\tau + \sigma \int_t^T e^{a\tau}dW_\tau \quad \Longrightarrow$$

$$X_T = e^{-a(T-t)}X_t + b\int_t^T e^{-a(T-\tau)}d\tau + \sigma e^{-aT}\int_t^T e^{a\tau}dW_\tau$$

Directly resolving the time integral this simplifies to:

$$X_T = X_t e^{-a(T-t)} + \frac{b}{a}\left(1 - e^{-a(T-t)}\right) + \sigma \int_t^T e^{-a(T-\tau)}dW_\tau$$

Using the properties (1.14) and (1.13) of the Ito Integral this reveals the distributional property of our process:

$$X_T \sim \mathcal{N}\left(X_t e^{-a(T-t)} + \frac{b}{a}\left(1 - e^{-a(T-t)}\right), \frac{\sigma^2}{2a}\left(1 - e^{-2a(T-t)}\right)\right)$$

$$\square$$

The above derivation would go similar for time-dependent parameters $b(t)$, $\sigma(t)$ and $a(t)$.

The process is said to have the property of *Mean-Reversion* with mean-reversion level $\frac{b}{a}$ and mean-reversion speed a. This has the explanation - assuming $a > 0$ that if X_t exceeds this level, the drift will become negative forcing the process to decrease (in expectation!). The analogue would apply for the reverse direction.

It remains to emphasize two important characteristics of the process (1.24):

$\diamond$ For $a > 0$ the process has expected finite long-time behavior:

$$\lim_{T\to\infty} \mathbb{E}[X_T|\mathcal{F}_t] = \frac{b}{a} \quad \text{and} \quad \lim_{T\to\infty} \text{Var}[X_T|\mathcal{F}_t] = \frac{\sigma^2}{2a}$$

$\diamond$ If the mean-reversion speed a tends to zero, we will have[13]:

$$\lim_{a\to 0} X_T \sim \mathcal{N}\left[X_t + b(T-t), \sigma^2(T-t)\right]$$

---

[13]This follows under the use of L'Hospital's Rule

1.6. Changing Probability Measures: Girsanov's Theorem

It will be important for later considerations to be able to change the statistics of the underlying path structure by switching from given probability measure P to another measure Q. In the current subsection the important Girsanov Theorem will be developed. The theorem states in the case of Brownian Motion with drift as in (1.4) that changing the drift is sufficient to change to a desired measure Q. In other words under different measures Brownian Motion becomes associated with different drift terms in the sense of (1.4).

Again consider the filtered Probability Space $(\Omega, \mathcal{F}, P; \{\mathcal{F}\}_t)$

Definition 1.13. *Two measures Q and P on $(\Omega, \mathcal{F})$ are said to be equivalent, if they define the same impossible events:*

$$P(A) = 0 \Leftrightarrow Q(A) = 0 \quad \forall A \in \mathcal{F}$$

The Radon-Nikodym Theorem - for a detailed treatment see [6, Chapter 28] - provides a sufficient and necessary condition for such an equivalent measure to exist. It states that given a probability measure P, an equivalent measure Q can be constructed, if there exists a random variable $Z(\omega) := \frac{dQ(\omega)}{dP(\omega)}$ in the sense that[14]:

$$Q(A) = \int_\Omega \mathbf{1}_A(\omega) dQ(\omega) = \int_\Omega \mathbf{1}_A(\omega) Z(\omega) dP(\omega) \quad \forall A \subset \mathcal{F} \quad (1.27)$$

Z has to be seen as a density and is termed the *Radon-Nikodym Derivative* of Q w.r.t P.

Girsanovs Theorem in its simplest form following [7, pp.191] relates Brownian Motions under different probability measures.

---

[14]Here $\mathbf{1}_A$ denotes the characteristic function: $\mathbf{1}_A(x) = 1$ if $x \in A$ and 0 otherwise

Theorem 1.3 (Girsanov). *Let there be a process M_t defined as an exponential of some adapted n-dimensional integrable process γ_t:*

$$M_t := \exp\Big(-\sum_{i=1}^{n}\int_0^t \gamma_\tau^i dW_\tau^{iP} - \frac{1}{2}\int_0^t \|\gamma_\tau\|^2 d\tau\Big) \tag{1.28}$$

Then M_t is a P-Martingale[15] and induces a new measure Q. The measure Q is equivalent to P and becomes constructed in the sense of (1.27) as[16]:

$$Q(A) := \int_\Omega \mathbf{1}_A \frac{M_T(\omega)}{M_t(\omega)} dP(\omega) = \mathbb{E}^P\Big[\mathbf{1}_A \frac{M_T}{M_t}\Big|\mathcal{F}_t\Big] \quad \forall A \in \mathcal{F}_t, \forall T \geq t \tag{1.29}$$

This measure Q is characterized by the property that a Ito Process defined as a P-Brownian Motion with drift (1.4):

$$dW_t^Q := dW_t^P + \gamma_t dt \tag{1.30}$$

becomes a Q-Brownian Motion - i.e. a driftless process.

Remark 1.10. *A remark on the proof: By applying Ito's Lemma (1.1) one verifies that M_t is the solution of the following driftless SDE - therefore a P-Martingale:*

$$dM_t = -\gamma_t M_t dW_t^P \tag{1.31}$$

This martingale transforms a measure P into a new measure Q under which a process X_t defined by the dynamics:

$$dX_t = \gamma_t dt + dW_t^P$$

will be driftless. This follows by an application of the Product Rule 1.2. The product process $Y_t := X_t M_t$ - i.e. the process X_t under the new measure Q - will be a martingale:

$$d[X_t M_t] = -X_t M_t \gamma_t dW_t^P + M_t\big[\gamma_t dt + dW_t^P\big] - \gamma_t M_t dt$$
$$= M_t[1 - X_t \gamma_t] dW_t^P$$

---

[15]Sufficient for M_t being a martingale using Definition 1.2 is the Nokidov Condition: $\exp(\frac{1}{2}\int_0^t \gamma^2 ds) < \infty$ (see [7, pp.198]) which we assume here to be satisfied.
[16]For the relation between Measure and Expectation: See also B.8, Appendix B

Remark 1.11. *The expectation of an adapted process X_t changes in the following manner:*

$$\mathbb{E}^Q[X_T|\mathcal{F}_t] = \mathbb{E}^P[X_T \frac{M_T}{M_t}|\mathcal{F}_t] \tag{1.32}$$

Changing from a measure P to another Q by multiplying with a P-Martingale can basically be interpreted as a change of the path probabilities. In other words a path under Q becomes more or less likely compared to the former measure P.

Although this simple version of the Theorem relates only Brownian Motions under different probability measures, it can be extended to any Ito Process having different drifts in each dimension which results in switching to other drifts by appropriately specifying γ_t.
For several extensions of the theorem for general Ito Processes the reader is referred to [4, pp.161]. For applications we refer to the following chapters on financial theory.

Remark 1.12 (Construction of a new Measure). *Applying the above Theorem 1.3 a new Measure is constructed in the following sense:*

1. *Start with an arbitrary SDE of the form: $dX_t^P = \mu_t dt + \sigma_t dW_t^P$*

2. *If (1.28) defined with $\gamma_t := \frac{\mu_t}{\sigma_t}$ is a martingale, then Q exists and with (1.30:*

$$dW_t^Q := dW_t^P + \frac{\mu_t}{\sigma_t}dt$$

3. *Result: $dX_t^Q = \sigma_t dW_t^Q$*

Remark 1.13. *The Girsanov Measure Transformation (1.30) does* not *change the volatility coefficient σ_t.*

1.7. Connection to PDEs: The Feynman-Kac Theorem

As already indicated in Section 1.2 the here introduced Feynman-Kac Theorem provides a remarkable concept within the framework of Markov processes and PDE's.

The Theorem states that the dynamics of the conditional expectation on a specific function $C(.)$, which depends on future outcomes of a process X_τ $(t \geq \tau)$ seen at a given system state $X_t = x$, satisfy a parabolic convection-diffusion PDE.

Necessary for that to happen is that the underlying stochastic process satisfies the Markov property (1.3) in the sense of SDE (1.20).

Furthermore the resulting PDE can be treated as a continuous time and space equivalent of the backward equation (1.5), which describes the time-discrete dynamics of conditional expectations backwards in time.

We provide the Theorem with the following specific function:

$$C(.) := \exp\Big[\int_t^T X_\tau d\tau\Big]\Phi(X_T)$$

inside the expectation operator[17] which depends on the outcomes X_τ at future time points $t \leq \tau \leq T$.

In the subsequent treatment we will see that in financial modeling this specific chosen function $C(.)$ is used to express the discounted value of a future cash flow $\Phi(X_T)$, where both the discount factor and the cash flow may depend on an underlying Markovian model system.

Theorem 1.4 (Feynman-Kac Theorem). *Let $X_t \in \mathbb{R}^n$ be a solution of the Markovian SDE (1.20) on $[0,T] \subseteq [0,T^*]$.*

Furthermore let there be $\Phi : \mathbb{R}^n \mapsto \mathbb{R}$, $r : [0,T] \times \mathbb{R}^n \mapsto \mathbb{R}$ integrable and $V(t,x) \in \mathcal{C}^1\big([0,T)\big) \times \mathcal{C}^2(\mathbb{R}^n)$ be defined by:

$$V := V(t,x) := \mathbb{E}\left[\exp\left[-\int_t^T r(\tau, X_\tau)d\tau \right] \Phi(X_T) \Big| X_t = x \right]$$

$$(1.33)$$

Then V solves the following PDE for $0 \le t < T$ and $x \in \mathbb{R}^n$:

$$\frac{\partial V}{\partial t} + \mathcal{L}V - rV = 0 \qquad (1.34)$$

with Final-Time Condition:

$$V(T,x) = \Phi(x) \qquad (1.35)$$

The Differential Operator $\mathcal{L}$ is called the Infinitesimal Generator *and is given by:*

$$\mathcal{L} = \sum_{j=1}^n a_i \frac{\partial}{\partial x_i} + \frac{1}{2} \sum_{i,j=1}^n <\sigma_i|\sigma^j> \frac{\partial^2}{\partial x_i \partial x_j} \qquad (1.36)$$

We give here a detailed proof through an application of Ito's Lemma 1.1.

Proof. We firstly introduce the stochastic process for $t \in [0,T]$ fixed:

$$B^t : [t,T] \times \mathbb{R}^n \mapsto \mathbb{R}$$

$$B_s^t := B^t(s, X_s) := \exp\left[-\int_t^s r(\tau, X_\tau)d\tau \right]$$

---

[17]There are also versions of this formula with other functions inside the expectation operator.

As B_s^t is a function on the process X_t we obtain from Ito's Lemma:

$$dB_s^t \bigg|_{s=t} = \left[\frac{\partial B^t}{\partial t} dt + \frac{\partial B^t}{\partial x} dX + \frac{1}{2} \frac{\partial^2 B^t}{\partial x^2} dX dX \right]_{s=t}$$

Evaluating the first partial derivative w.r.t. the process X:

$$\frac{\partial B^t}{\partial x} \bigg|_{s=t} = -B_s^t \int_t^s \frac{\partial r}{\partial x} d\tau \bigg|_{s=t} = 0$$

The same applies for the second derivative so we are left with:

$$dB_s^t \bigg|_{s=t} = \frac{\partial B^t}{\partial t} dt \bigg|_{s=t} = -r B_s^t |_{s=t} = -r B_t^t = -r$$

The proof is done now in three steps.

⋄ Firstly note that the construction of $V(t,x)$ as in (1.33) works only, when X_t satisfies the Markov Property (1.3). As we have shown in Proposition 1.1 a stochastic process V_t can be constructed as the conditional expectation of a function depending on the future outcome of the process X_τ with $\tau \geq t$. We firstly define the process V_t as:

$$V_t := \mathbb{E}[B_T^t \Phi(X_T) | \mathcal{F}_t]$$

This process as a conditional expectation is itself a martingale measurable w.r.t $\mathcal{F}_t$. As the process X_t in the above theorem satisfies the Markov Property, $\mathcal{F}_t$ simplifies to the σ-algebra generated by X_t alone - i.e. $\sigma(X_t)$. Therefore V_t can be parameterized as a function depending on X_t alone:

$$V(t, X_t) := V_t = \mathbb{E}[B_T^t \Phi(X_T) | \sigma(X_t)]$$

⋄ Secondly we take a look on the final-time condition (1.35), which we obtain directly with the above defined process V_t:

$$V(T, X_T) \overset{1.33}{=} \mathbb{E}[B_T^T \Phi(X_T) | \sigma(X_T)] = \Phi(X_T)$$

For a specific outcome $X_T = x$ we have $V(T, x) = \Phi(x)$.

$\diamond$ To derive the PDE in the last step we now introduce the product process:

$$U_s := U(s, X_s) := B_s^t V(s, X_s), \quad s \in [t, T]$$

As U_s depends on the Markovian Process X_t this allows us to apply Ito's Lemma - in particular the Product Rule 1.2 - which writes in short-hand notation in combination with the properties of the process $B^{t\,18}$:

$$dU_s\Big|_{s=t} = B_s^t dV_s + V(t, X_t)dB_s + \underbrace{dV_s dB_s}_{(1)}\Big|_{s=t}$$

$$= V_t dt + V_x dX_t + \frac{1}{2}V_{xx}dX_t dX_t - rV_t$$

$$= \Big[V_t + a^T V_x + \frac{1}{2}[\sigma^T \sigma]V_{xx} - rV\Big]dt + V_x \sigma dW_t$$

$$= \Big[V_t + \mathcal{L}V - rV\Big]dt + V_x \sigma dW_t$$

We do not have to consider (1), since this term results in terms of higher asymptotic order $O(\Delta X \Delta t)$ which can be neglected in the sense of Ito's Formula. Now integration yields:

$$U(T, X_T) = B_T^t V(T, X_T) = B_T^t \Phi(X_T)$$

$$= \underbrace{U(t, X_t)}_{=V(t,X_t)} + \int_t^T \Big[V_t + \mathcal{L}V - rV\Big]d\tau + \int_t^T V_x \sigma dW_\tau$$

Under conditional expectation with a specific outcome $X_t = x$ this writes:

$$\mathbb{E}[B_T^t \Phi(X_T)|X_t = x] = V(t, x)$$

$$+ \mathbb{E}[\int_t^T \Big[V_t + \mathcal{L}V - rV\Big]d\tau|X_t = x] + \underbrace{\mathbb{E}[\int_t^T V_x \sigma dW_\tau|X_t = x]}_{=0}$$

From the definition of $V(t, x)$ in (1.33) and the conditional expectation of $U(T, X_T)$ being itself a martingale (Proposition 1.1) this *necessarily* implies:

$$V_t + \mathcal{L}V - rV = 0$$

$\square$

Note the necessity for X_t to satisfy the Markov property. In an e.g. Non-Markovian case (see Remark 1.8) we could not have simplified the filtration $\mathcal{F}_t$ to the σ-Algebra generated by X_t alone and could not have parameterized the product process U_t to apply Ito's Lemma.

---

[18] Here a^T denotes the transpose of a vector a

1.8. Applications in Finance

Prices of *Tradeable Securities* get modeled as stochastic processes $S(t,\omega)$ on an arbitrary filtered probability space $(\Omega, \Sigma, P; \{\mathcal{F}\})$ satisfying SDE's of the form:

$$dS_t = \mu_t(t, S_t)dt + \sigma_t(t, S_t)dW_t^P \tag{1.37}$$

with arbitrary *expected return* μ and - as a quantification of riskiness - *volatility* coefficients σ as introduced in Definition 1.12. Due to the upcoming of *Derivative Securities*, whose future payoff $\Phi(S_T)$ is based on the performance on an underlying security S_T at a future time point T, the question arose, what rational value V_t had to be assigned to those also termed *Contigent Claims* in the time prior to their final payoff.

The main essence due to F. Black and M. Scholes in their famous paper[19] is that two main conditions are sufficient to model the value of contingent claims in an economically rational way, particularly independent of the investors individual expected returns μ. In the modeling context this means that any model which is built to value a contingent claim in a security market has to be tested on those conditions. For a detailed discussion we refer to [8, pp.80].

Absence of Arbitrage

The first important condition is that it is not possible to generate a riskless profit called *Arbitrage*. In such an arbitrage-free market the best prediction on the future value of a security an investor can make, is the currently known one. In modeling terms this means that a probability measure has to be found, under which the assumed stochastic process is a martingale. Furthermore to make the price of all traded securities and cash flows of contingent claims at different time points comparable, they are valued w.r.t to one *reference unit* also termed the *Numeraire*. In the Black-Scholes setting this is done in terms of a risk-free cash account, which is continuously compounded by a constant risk-free rate r.

---

[19]The Pricing of Options and Corporate Liabilities. Journal of Political Economy, 81 (1973), 637-654

This numeraire is represented here through the *Discount Factor*[20]:

$$D(t,T) := \exp[-r(T-t)] \tag{1.38}$$

Multiplication of this factor with a specific future payoff $\Phi(S_T)$ at time T yields its *Present Value* at time $t < T$.

Girsanov's Theorem 1.3 now tells how and especially when one is able to switch from an arbitrary measure P to such a *Martingale Pricing Measure*, usually termed the *Risk-Neutral Measure* (RN), where now all *discounted* securities $D(s,t)S_t$ ($\forall s \leq t$) are martingales, or equivalently that the risk-neutral price process S_t^{RN} solves the measure-transformed SDE:

$$dS_t^{RN} = rS_t^{RN}dt + \sigma(t, S_t^{RN})dW_t^{RN} \quad \text{with} \tag{1.39}$$
$$dW_t^{RN} := dW_t^P + \lambda_t dt$$

The process $\lambda(t,\omega)$ defines the exponential martingale (1.28). In this particular context (1.37) the process would have to become defined by:

$$\lambda_t(\omega) := \frac{\mu_t - r}{\sigma_t} \tag{1.40}$$

which is known as the *Market Price of Risk* as the excess return over the risk-free rate measured in standard deviations. For a further discussion of that specific process we refer to [8, pp.115].

Theorem 1.5 (Risk-Neutral Valuation). *Under the* Risk-Neutral Measure *the discounted value at time t V_t of a cash-flow $\Phi(.) : \mathbb{R} \to \mathbb{R}$ depending on the future value of an underlying security S_T at $T \geq t$ is given by:*

$$\boxed{V_t = \mathbb{E}^{RN}\left[D(t,T)\Phi(S_T)\big|\mathcal{F}_t\right] := \mathbb{E}\left[D(t,T)\Phi(S_T^{RN})\big|\mathcal{F}_t\right]} \tag{1.41}$$

Remark 1.14. *The risk-neutral value process V_t is itself a martingale with the construction as in Proposition 1.1.*

---

[20]This is sometimes reversely (more intuitively) defined by the notion of a Money-Market Account B_t, whose time evolution is described by the ODE: $dB = rBdt$

Market Completeness

Since now having defined V_t in mathematical terms but working at a real market implies that its value should be expressed in terms of quantities available in the market. Setting up a portfolio to evaluate V_t is called *Replication* in the sense that

$$V_t = \phi_t S_t + \psi_t B_t \qquad (1.42)$$

with amount ϕ_t of the underlying security S_t and an amount of ψ_t of cash units B_t at time t. To model the evolution of that value only in terms of the chosen securities one postulates the *Self-Financing* Condition

$$dV_t = \phi_t dS_t + \psi_t dB_t \qquad (1.43)$$

This implies that any change in the value of a contingent claim dV_t can be set to zero by setting up a suitable *Portfolio Strategy* (ϕ_t, ψ_t) against it. This elimination of the risk of unexpected changes in the value is termed *Hedging*.

The second condition of *Market Completeness* now postulates that at any time point the value V_t can be uniquely replicated in the sense of (1.43). Especially the Black-Scholes Model satisfies that second condition but modeling the securities in a different way results in an *incomplete market*, where it is not possible to eliminate risk at any time point. The important conclusion is that if there is more than one equivalent Risk-Neutral Measure then the market will be incomplete (see e.g. [8, pp.196] for more details).

An overview of the Black-Scholes Model and the derivation of the corresponding risk-neutral valuation PDE is given in Appendix A.1.

Valuation in Arbitrary Units: The Change-of-Numeraire Technique

Usually - as in the Risk-Neutral Pricing Theorem 1.5 - the reference asset is chosen to be the risk-free cash account respectively the discount factor $D(t, T)$. Beyond that a risk-neutral valuation can be generalized to be done with respect to any tradeable security.

Definition 1.14 (Numeraire). *A Numeraire $N : [0, T^*] \times \Omega \to (0, \infty)$ is a tradeable security modeled as a positive adapted stochastic process with martingale dynamics under the risk-neutral measure.*

As a tradeable security the numeraire has to be arbitrage free under the risk-neutral measure. Valuation of securities and contingent claims under arbitrary numeraires implies a change of measure such that under that new measure every tradeable security relative to the new numeraire will again be a martingale. In particular it is proven that to every arbitrary numeraire N_t there exists a equivalent martingale measure $\mathbb{Q}^N$. This result can be found e.g. in [9, p.27]. We explain here in a more informal way how this can be done.

Theorem 1.6 (Numeraire-Change). *Let there be N_t and U_t two numeraires under the martingale pricing-measure $\mathbb{Q}^N$ associated with numeraire N_t - for instance the risk-neutral measure with the risk-free cash account as in Theorem 1.5.*

$$\frac{V_t}{N_t} = \mathbb{E}^N\left[\frac{\Phi_T}{N_T}\Big|\mathcal{F}_t\right]$$

Multiplying both sides with the $\mathcal{F}_t$-adapted process $\frac{N_t}{U_t}$ - which can be dragged inside the $\mathcal{F}_t$-conditional expectation since it is $\mathcal{F}_t$ measurable[21] - and extending the fraction inside the expectation yields as in Remark 1.11:

$$\frac{V_t}{N_t}\frac{N_t}{U_t} = \frac{V_t}{U_t} = E^N\left[\frac{\Phi_T}{N_T}\frac{N_t}{U_t}\Big|\mathcal{F}_t\right] = \mathbb{E}^N\left[\frac{\Phi_T}{U_T}\underbrace{\frac{U_T N_t}{N_T U_t}}_{(1)}\Big|\mathcal{F}_t\right] =: \mathbb{E}^U\left[\frac{\Phi_T}{U_T}\Big|\mathcal{F}_t\right]$$

$$(1.44)$$

Here (1) *denotes the Radon-Nikodym Derivative (1.29) in this context which is responsible for the particular measure change.*

As we are assuming any tradeable security to follow an Ito Process, we can also interpret that result in view of Girsanovs Theorem 1.3. There we have shown that a *change of measure* basically implies a *change of the drift* of the underlying Ito Process dynamics. Especially since N_t and U_t are martingales under $\mathbb{Q}^N$, fraction (1) becomes exactly the exponential martingale (1.28) which is used in Girsanov's Theorem (1.3) to express Brownian Motions under different measures.
To see this consider the following simplified example assuming driftless lognormal dynamics (1.21) and constant coefficients for N_t and U_t.

Example 1.1.

$$dN_t = \sigma_N N_t dW_t^N \quad and \quad dU_t = \sigma_U U_t dW_t^N$$

Since changing from numeraire N_t to U_t (respectively changing the measure and therefore the path statistics) we can directly calculate the relationship of Brownian Motions under both numeraires by applying the Ito Product Rule 1.2:

$$d\left[\frac{N_t}{U_t}\right] = \frac{1}{U_t}dN_t + N_t d\frac{1}{U_t} + dNd\frac{1}{U_t}$$

$$= \frac{N_t}{U_t}\sigma_N dW_t^N + N_t\left[-\frac{1}{U_t^2}\sigma_U U_t dW_t^N\right.$$

$$+ \left.\frac{1}{U_t^3}\sigma_U^2 U_t^2 dW_t^N\right] - \frac{N_t}{U_t}\sigma_N\sigma_u dt$$

$$= \frac{N_t}{U_t}\left[\left[\sigma_N - \sigma_U\right]dW_t^N - \left[\sigma_N - \sigma_U\right]\sigma_U dt\right]$$

To have martingale dynamics under the new numeraire U_t we necessarily have to set in the sense of (1.30):

$$dW_t^U := dW_t^N - \sigma_U dt \qquad (1.45)$$

---

[21]For that, see also the properties of conditional expectation in Appendix B

*This also defines our new measure $\mathbb{Q}^U$ as we did with formula (1.29).
Integration of the last result in the sense of (1.21) leads exactly to frac-
tion (1) which is then the exponential martingale used in Girsanov's
Theorem in (1.28):*

$$\int_t^T d\log\left[\frac{N_\tau}{U_\tau}\right] = \frac{N_T U_t}{N_t U_T}$$

$$= \exp\left[\int_t^T [\sigma_N - \sigma_U] dW_\tau^N - \frac{1}{2}\int_t^T [\sigma_N - \sigma_U]^2 d\tau\right]$$

A derivation of how the drift changes for more general SDE's when
switching to a different numeraire can be found in [9, p.30].

2. Fixed Income Markets

> *If you want to know the value of a security, use the price of another security that is as similar to it as possible. All the rest is modeling.*
>
> Emanuel Derman

We proceed considering the market, whose products depend more or less[1] exclusively on interest rate quantities. These are usually termed *Fixed Income Securities* and mainly represented by *Bonds* which are issued to cover financial needs of banks, companies or the government in terms of external financing.

We firstly describe the fixed income market and introduce its main traded contracts from a practical perspective. A general modeling approach to value interest rate dependent securities under the influence of stochastic interest rates will be reviewed in the subsequent section. For other treatments we recommend [9, chs. 1 and 2] and [10, ch. 9].

---

[1]We assume absence of any credit risk

2.1. The Yield Curve

Definition 2.1 (Zero-Coupon Bond). *A Zero-Coupon Bond*

$$P(\cdot, T) : [0, T] \to (0, \infty) \quad \text{with Maturity } T \in [0, T^*] \tag{2.1}$$

is a tradeable security, which guarantees its holder the payment of one unit of money at maturity: $P(T, T) = 1$. *Its present value at time* $t \leq T$ *is denoted as* $P(t, T)$. *The bond's value at time* $t \in [0, T^*]$ *as a function w.r.t maturity*

$$T \mapsto P(t, T) \quad \forall \quad T \geq t \tag{2.2}$$

is known as the Zero-Bond Curve *or* Discount Curve.

There are now several ways to derive interest rate quantities out of given bond values $P(t, T)$, which are taken from market quotes.

Definition 2.2 (Yields and Term Rates). *Assuming* Continuously Compounded *interest rates termed as* Yields $Y(t, T)$ *the value becomes:*

$$P(t, T) := \exp\left[-Y(t, T)(T - t)\right] \Rightarrow Y(t, T) := -\frac{\log P(t, T)}{T - t}$$
$$\tag{2.3}$$

Simple Compounding *yields the* Term Rate $L(t, T)$ *with Tenor* $T - t$:

$$P(t, T) := \frac{1}{1 + L(t, T)(T - t)} \Rightarrow L(t, T) := \frac{1 - P(t, T)}{(T - t)P(t, T)} \tag{2.4}$$

which also leads directly to a definition of Future *Term Rates* $L(T_1, T_2)$ *for simple compounding starting at a future start-time point* $T_1 > t$ *and with Tenor* $T_2 - T_1$:

$$L(T_1, T_2) = \frac{1 - P(T_1, T_2)}{(T_2 - T_1)P(T_1, T_2)} \tag{2.5}$$

As an important feature - contrarily to yields - some term rates are directly provided in the market such as LIBOR rates (**L**ondon **I**nterbank **O**ffered **R**ate). The Libor Rates are reference rates quoted for short term maturities up to one year. They represent *Interbank Rates* at which banks are lending at and borrowing from each other.

Definition 2.3 (Short Rate). *Assuming smoothness of (2.2) and taking the limit yields the* Short Rate*:*

$$r(t) := \lim_{T \searrow t} Y(t, T) = \lim_{T \searrow t} L(t, T) \tag{2.6}$$

This quantity represents - a rather theoretically idealized - current interest amount one has to pay when borrowing over an infinitesimal small interval of time which is also known as Instantaneous Borrowing.
In turn we have a time-dependent generalization of the discount factor (1.38) as the money unit numeraire:

$$D(t, T) = \exp\left[-\int_t^T r(\tau) d\tau \right] \tag{2.7}$$

Forward Rates and the Yield Curve

Starting from a current zero-bond curve (2.2) we can also derive the current interest rate estimated by the market for borrowing at future time points $T_i > t$ termed as *Forward Rates*. They describe the amount one has to invest at a future time $T_1 > t$ to receive one unit at time $T_2 > T_1$. As seen from the current time t this yields a relationship between the bonds $P(t, T_1)$ and $P(t, T_2)$, which is given in a simple compounded environment (2.4) as:

$$P(t, T_2)(1 + K)(T_2 - T_1) = P(t, T_1) \tag{2.8}$$

Solving this equation for K yields a quantity, which implicitly determined in currently (i.e. at time t) available market quotes. It can be interpreted as the *best prediction* of a future term rate $L(T_1, T_2)$, which the market can currently make. This interest rate quantity is known as the *Forward Rate*.

Definition 2.4 (Forward Rate). *The* Forward Term Rate *of forward borrowing at time T_1 with maturity $T_2 > T_1$ and $\Delta T = T_2 - T_1$ is given by:*

$$F(t; T_1, T_2) := \frac{1}{\Delta T} \frac{P(t, T_1) - P(t, T_2)}{P(t, T_2)}, \quad \forall T \geq t \qquad (2.9)$$

Assuming smoothness of (2.2) and taking the limit yields with $T := T_1$ and $T + \Delta T := T_2$ the Forward Rate of Instantaneous Borrowing *at time T -* Forward Rate *for short[2]:*

$$f(t, T) := \lim_{\Delta T \to 0} F(t; T, T + \Delta T) = -\frac{\partial \log P(t, T)}{\partial T}, \quad \forall T \geq t$$
$$(2.10)$$

In turn the bond price becomes in terms of forward rates:

$$P(t, T) = \exp\left[-\int_t^T f(t, \tau) d\tau\right] \qquad (2.11)$$

Definition 2.5 (Yield Curve). *At any given time $t \in [0, T^*]$ forward rates as a function of maturity are known as the* Yield Curve.

$$T \mapsto f(t, T), \quad \forall T \geq t \qquad (2.12)$$

This curve shows the amount of interest as the price for instantaneous borrowing at every future time point $T \geq t$.

Instantaneous forward rates are rather a mathematical idealization, since it requires the discount curve (2.2) to be a continuous and differentiable function, although only a countable set of maturities is observable in the market. Therefore numerical interpolation schemes have to be applied. For a mainly used algorithm called *Bootstrapping*, which extracts the discount curve out of the prices of several interest-rate market securities, we refer to [11, pp.146].

---

[2]Note that we have here the partial derivative w.r.t Maturity T instead of Time t

Remark 2.1. *The short rate (2.6) is one element of the yield curve, since we have:*

$$r(t) := f(t,T)\Big|_{T=t} := -\frac{\partial \log P(t,T)}{\partial T}\Big|_{T=t} \tag{2.13}$$

Remark 2.2. *With formula (2.9) a forward rate can be interpreted as a tradeable security, since it is simply the multiple $\frac{1}{\Delta T}$ of the difference of two zero coupon bonds $P(t,T_1)$ and $P(t,T_2)$ valued in units of $P(t,T_2)$.*

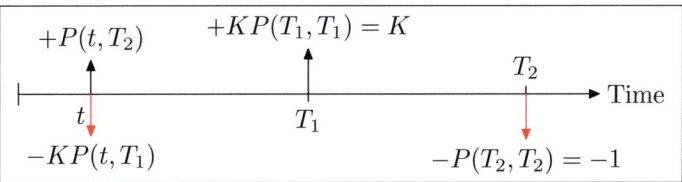

Figure 2.1.: A Forward Term-Rate represented $F(t,T_1,T_2)$ as a Zero-
Coupon Bond Portfolio

Figure 2.1 shows the construction of the arbitrage-free price K at time t for forward borrowing from a future time point T_1 with maturity T_2 by setting up a portfolio in buying one unit of $P(t,T_2)$ and selling k units of $P(t,T_1)$. Setting the portfolio value to zero at time t yields for the no-arbitrage price: $K = \frac{P(t,T_2)}{P(t,T_1)}$. With an expression in terms of an interest rate quantity one arrives exactly at the Forward Term Rate as in (2.8).

2.2. Interest Rate Securities

A zero-coupon bond is just one traded security in the fixed income market. In this section we describe Coupon Bearing Bonds and Interest Rate Swaps, which are fundamental for derivative products described in Section 2.4.

Their market values are given in terms of forward rates (2.9). This can also be made consistent from a modeling point of view when these quantities get modeled under stochastic interest rates (in particular see Remark 2.3). The resulting valuation done in bond units under the Forward Measure is reviewed in the subsequent section.
Firstly it is useful to introduce the following notation.

Tenor Structure

The Tenor Structure $\mathcal{T}$ of an interest rate product to be valued at time $t \in [0, T^*]$ is a set of not necessarily equally distributed time points T_i.

$$\mathcal{T} = \{T_0, T_1, \ldots, T_N\} \quad \text{where} \quad t \leq T_0 < T_{i+1} \quad \forall i = 0, \ldots, N-1$$
$$\Delta_i := T_i - T_{i-1}, \ i = 1, \ldots, N$$

The Tenor Length is the length of the time interval $T_N - T_0$.

Coupon Bonds (CB)

When we have to deal with coupon bearing bonds, we receive the interest rate earnings as coupon payments - denoted as C_i - at tenor dates T_i. As the interest rate R is fixed here, this specifies the i-th coupon value to be $C_i = \Delta_i R$. The present value of a coupon bond over a tenor $\mathcal{T}$ is given as the discounted sum of all future coupon payments and the final repayment of the notional amount, which is here assumed to be one unit.

$$\mathrm{CB}(t, T_N) = \sum_{i=1}^{N} P(t, T_i)C_i + P(t, T_b) \tag{2.14}$$

Floating Rate Notes (FRN)

If the interest rate R is not fixed as above, it will be usually reset after each tenor date T_i in the sense that it will be readjusted (i.e. fixed) to be equal to the current market rate (usually the LIBOR rate) after each coupon payment. This means that we have here fixing dates $T_0 \ldots T_{N-1}$ and payment or coupon dates $T_1 \ldots T_N$. Therefore the value of such a bond valued in bond units is always equal to its notional amount at the first tenor date T_0. When using the coupon-bond formula (2.14) in combination with the definition of forward term rates (2.9) and the final repayment we have at value time $t \leq T_0$:

$$
\begin{aligned}
\mathrm{FRN}(t, T_N) &= \sum_{i=1}^{N} P(t, T_i) \Delta_i F(t, T_{i-1}, T_i) + P(t, T_b) \\
&\overset{2.9}{=} \sum_{i=1}^{N} \left[P(t, T_{i-1}) - P(t, T_i) \right] + P(t, T_b) \\
&= P(t, T_0)
\end{aligned}
\tag{2.15}
$$

Interest Rate Swaps (IRS) and the Swap Rate

A prototypical forward starting IRS is a *contract*, which exchanges a floating-rate based against a fixed-rate based cash flow based on a nominal amount of one unit with a given tenor structure $\mathcal{T}$ starting at a future date $T_0 \geq t$.

Usually the two components of a swap are referred to as *fixed and floating legs*. The fixed leg can be valued according to a Coupon Bond where the floating side is equal to a Floating Rate Note each of both *without* final repayments. Buying a *Payer IRS* (PRS) the holder receives floating and pays fixed rates, where buying a *Receiver IRS* (RFS) he will do the reverse.

Exemplarily we obtain the swap value for a PRS:

$$
\begin{aligned}
\mathrm{PRS}(t; \mathcal{T}, R) &= \sum_{i=1}^{N} P(t, T_i) \Delta_i \big[F(t, T_{i-1}, T_i) - R \big] \\
&\stackrel{2.9}{=} \sum_{i=1}^{N} \big[P(t, T_{i-1}) - P(t, T_i) \big] - \sum_{i=1}^{N} P(t, T_i) \Delta_i R \\
&= P(t, T_0) - P(t, T_N) - \sum_{i=1}^{N} P(t, T_i) \Delta_i R \qquad (2.16)
\end{aligned}
$$

Given a fixed swap structure we obtain the no-arbitrage value of this product by setting the PRS-Value to zero and solve for the fixed rate R. This quantity is known as the *Par Swap Rate* :

$$
S_{0,N} := R = \frac{P(t, T_0) - P(t, T_N)}{\sum_{i=1}^{N} \Delta_i P(t, T_i)} \qquad (2.17)
$$

2.3. Interest Rate Derivatives

After having discussed the major traded securities we now proceed addressing the derivatives market, where products are traded whose future payoff depends on the performance of underlying securities introduced above. This market is basically classified by three main product forms:

⋄ Bond Options

⋄ Options on Interest Rates termed as *Caps and Floors*

⋄ Options on Swaps termed as *Swaptions*

We shortly introduce those derivatives and also emphasize how they are associated with each other.

Bond Options

These derivatives do not differ from standard equity options except that the underlying is now a zero coupon respectively a coupon bond. Exemplarily for a zero-coupon bond call (ZBC) with bond maturity T_2 and option maturity T_1 (surely $T_1 < T_2$) and with strike K the payoff is[3]:

$$\Phi_{ZBC}[P(T_1, T_2)] := \big[P(T_1, T_2) - K\big]_+ \qquad (2.18)$$

Equivalently for a coupon bond option call (CBC) we have with formula (2.14):

$$\Phi_{CBC}[CB(t, T_2)] = \big[CB(t, T_2) - K\big]_+ \qquad (2.19)$$

$$= \Big[\sum_{i=1}^{N} P(t, T_i)C_i - K\Big]_+ \text{ with } C_i = \Delta_i R \quad (2.20)$$

As a bond price is a monotone decreasing function of interest rates, we can also interpret those derivatives as options on interest rates (see Appendices A.3 and A.4).

---

[3] $\big[f(x)\big]_+$ is a common writing for $\max\big[f(x), 0\big]$ in the financial literature

Caps and Floors

A European Caplet protects against increasing interest rates ("caps" the interest rate), and pays the difference between the market term rate $L(T_{i-1}, T_i)$ and a specified strike rate K at a maturity date T_{i-1}, if the difference is positive. A prototypical Caplet with notional amount $N = 1$ corresponds to a call option on the future term rate. The Payoff of a Caplet at time T_i on a Future Term Rate $L_i := L(T_{i-1}, T_i)$ with maturity (i.e. fixing time) T_{i-1} and tenor $T_i - T_{i-1}$ is given by:

$$\Phi_{\text{Caplet}}(L_i) = \Delta_i \big[L_i - K \big]_+ \qquad (2.21)$$

Respectively for the discounted payoff at time T_{i-1}:

$$D(t, T_i)\Phi_{\text{Caplet}}(L_i) = D(t, T_i)\Delta_i \big[L_i - K \big]_+ \qquad (2.22)$$

A sum of caplets with a specific tenor structure $\mathcal{T}$ is termed *Cap*. In analogy a Floorlet (Floor) corresponds to a (sum of) put option(s) on term rate(s). Furthermore a caplet (floorlet) can be expressed as a portfolio consisting of zero-coupon bond puts (calls) (see Appendix A.3).

Binary/Digital Caplets

The payoff of a Digital Caplet at time T_i with maturity T_{i-1} and with strike rate K on a Term Rate $L(T_{i-1}, T_i)$ is defined by:

$$\Phi_{\text{DigitalCaplet}}(L_i) = \begin{cases} 1, & L(T_{i-1}, T_i) > K \\ 0, & L(T_{i-1}, T_i) \le K \end{cases}$$

It can be shown that the expected value of a Digital Caplet is the expected value of a Caplet partially derived after the strike rate K. Formally:

$$\mathbb{E}[\Phi_{\text{DigitalCaplet}}(L_i)] = -\frac{\partial \mathbb{E}[\Phi_{\text{Caplet}}(L_i)]}{\partial K} \qquad (2.23)$$

For a derivation of this result we refer to [11, p.159].

European Swaptions

An European Swaption gives its owner the right to enter into a (Payer/Receiver) Interest Rate Swap at a specified exercise date $T_E \leq T_0$ where T_0 denotes the tenor start date of the underlying swap contract. The payoff of a Payer Swaption at T_E in combination with formula (2.16) is given as:

$$\Phi_{\text{Swaption}} = \left[\sum_{i=1}^{N} P(T_E, T_i) \Delta_i \left[F(T_E, T_{i-1}, T_i) - R \right] \right]_+$$

$$\stackrel{2.16}{=} \left[P(T_E, T_0) - P(T_E, T_N) - \sum_{i=1}^{N} P(T_E, T_i) \Delta_i R \right]_+$$

$$(2.24)$$

Dividing by the above sum we can see more directly, that a Swaption is an option on a future Swap-Rate at time T_E:

$$\Phi_{\text{Swaption}} = \left[\frac{P(T_E, T_a) - P(T_E, T_b)}{\sum_{i=1}^{N} P(T_E, T_i) \Delta_i} - R \right]_+$$

Under specific assumptions we can express the payoff of a swaption as a sum of bond options - see Appendix A.4. The market practice to value such standard or plain derivative products requires some more explanation and will be done by the introduction of Black's Formula.

Exotic Derivatives and Modeling

The above introduced prototypical interest rate derivatives are usually called *European* or *Plain* derivatives and are just one small segment in the entire derivatives market. To cope with the risk resulting from more individual designed underlyings, the complexity of corresponding derivatives, which are issued on those securities, is also increasing. This justifies the need for a sophisticated interest rate valuation model.

Such a model should be redundant for plain derivatives, since their prices are already provided in a liquid market environment. With increasing complexity and individuality of a derivative product, the market for such specific structures tends to become less liquid.

This can be augmented, until due to the product individuality a quoted market price would not be available anymore. That circumstance forces the pricing to be done by the use of a mathematical valuation model.

A model to value such derivatives gets firstly calibrated to the known (plain) market segment such that it reproduces the prices from the plain derivatives market. It is used afterwards in an extrapolating manner to predict a close market approximation for the specific exotic structure. For a good review on exotic interest derivatives we refer to [11, pp.169]. Here we would like to present exemplarily an early-exercise product.

Early Exercise Products

Early Exercise Derivative Products enable the owner to execute the right being inherent in the derivative payoff (such as an option) not at only one point in time (as usually in the European Case) but at several time-points. If any time point over the live time of a derivative is possible, these products are termed *American* and if there is only a specified set of discrete time points, they are named *Bermudan*. Due to that larger flexibility for the holder their values should be assumed to be always larger than the corresponding European Product. We now give a product description of an early exercise interest rate derivative, whose corresponding prototypical analogue has already been discussed in the case of the European swaption.

Bermudan Swaption

A Bermudan Swaption gives its owner the right but not the obligation to enter into an Interest Rate Swap at any tenor point of its associated tenor structure. Let there be a tenor structure $\mathcal{T} = \{T_0, \ldots, T_N\}$ of an underlying swap (payer or receiver) as introduced above. The payoff - respectively value V - at the tenor points T_i of a Bermudan Swaption is recursively defined by the following law:

$$V_{\text{BSwpt}}(T_N) = 0$$

$$V_{\text{BSwpt}}(T_i) = \max\left[V_{\text{Swap}}(T_i; T_i), V_{\text{BSwpt}}(T_i; T_{i+1})\right] \quad (i = N-1, \ldots, 0)$$

where we make use of the following notations:
The swap value of the underlying (remaining) payer or receiver swap at time T_i with tenor structure $T_i, \ldots, T_n$ is denoted as:

$$V_{\text{Swap}}(T_i; T_i) := \text{PRS/RFS}(T_i; \{T_i, \ldots, T_n\}, R)$$

where for the value of the Bermudan Swaption at time T_i starting at time T_{i+1} we write:

$$V_{\text{BSwpt}}(T_i; T_{i+1}) = V_{\text{BSwpt}}(T_i; T_{i+1}, \ldots, T_n)$$

2.4. General Modeling Approach

In contrast to stocks the values of cash flows of each fixed income instrument are sensitive to movements of *one* quantity: The Yield Curve (2.12). This curve is itself driven by movements of interest rates at several time points which marks the starting point for a modeling approach.

Basically the short rate (2.6) is now assumed to follow a stochastic process $r = r_t(\omega)$. This naturally implies for the discount factor (2.7) to become redefined in a third version as now also being a random variable:

$$D(t,T) := D_t(\omega, T) = \exp\left[-\int_t^T r_\tau(\omega)d\tau\right] \quad \forall \omega \in \Omega \qquad (2.25)$$

This marks a big distinction to the Black-Scholes equity model (Appendix A.1), since we augment the abstraction level by modeling a quantity, which itself does *not* have a tradeable character - we cannot buy an interest rate in the market - but on which the present value of any interest rate security depends on.

From now on we simply assume absence of any arbitrage between bonds $P(t, T_i)$ with different maturities T_i.

Theorem 2.1 (Risk-Neutral Bond Valuation). *Under the risk-neutral measure the model zero-bond price $P(t,T)$[4] by:*

$$\boxed{P(t,T) = \mathbb{E}^{RN}[D(t,T)|\mathcal{F}_t]} \qquad (2.26)$$

$P(t,T)$ is a stochastic process as a function depending on the actual and future outcomes of the short rate r_τ with $t \leq \tau \leq T$.

Proof. Use formula (1.41) together with $\Phi_T = P(T,T) = 1$ to obtain (2.26). $P(t,T)$ as the risk-neutral conditional expectation of the discount factor (2.25) becomes itself a stochastic process under the use of Proposition 1.1. $\qquad \square$

---

[4]Using the Tower Law of conditional probability we would also know how to price $P(T_1, T_2)$ at time $t \leq T_1 \leq T_2$: $\mathbb{E}^{RN}[D(t,T_1)P(T_1,T_2)|\mathcal{F}_t] = \mathbb{E}^{RN}[D(t,T_1)\mathbb{E}^{RN}[D(T_1,T_2)|\mathcal{F}_{T_1}]|\mathcal{F}_t]$

The risk-neutral dynamics of the stochastic zero-coupon bond (2.26) are given with an arbitrary volatility coefficient $\sigma_P(t,T)$ by[5]:

$$dP(t,T) = r_t P(t,T)dt + \sigma_P(t,T)dW_t^{RN} \tag{2.27}$$

Consistently this implies that the dynamics of the discounted bond price $D(s,t)P(t,T)$ for $s \leq t$ satisfies the martingale property (1.2).
Contrarily the non-tradeable character of the short rate itself leaves us much freedom to model this quantity in terms of an arbitrary - usually Markovian - Ito process in the sense of (1.20).
For more details on the derivation of the SDE (2.27) and the above applied change of measure we refer to [10, pp.331].

There are basically two main modeling approaches:

$\diamond$ The first one defines the *Short Rate* (2.6) to follow an Ito Process usually assuming Markovian dynamics with mean-reversion and therefore similar to (1.24). This has shown to be convenient, since due to the exactly known probability distribution we can calculate analytically the expected value of the bond price using above formula (2.26). As we then have the calculated *Model Bond Price* we can derive model prices of other quantities as introduced in Section 2.1. From that we are able to analytically calculate model values for plain derivative products.

$\diamond$ The second approach *directly* models families of *forward rates* $\{f(t,T)\}_{T \geq t}$ or term rates and therefore the entire yield curve exogenously as multidimensional Ito Processes and uses formula (2.11) to price bonds. In contrast to the short-rate model approach we do not have to know the model bond price to derive model prices for other quantities.

These two approaches are further discussed in the Chapter 3.

---

[5]If we for instance specify $\sigma_P(t,T) := \sigma_P P(t,T)$ we will obtain lognormal dynamics (1.21)

The Forward Measure

To value arbitrary contingent interest rate claims $\Phi(.)$ with the risk-neutral formula (1.41) we would have to calculate the expectation with the stochastic discount factor (2.25) inside. To get to an analytical solution this would mean to know the joint density between the random variables $\Phi(.)$ and $D(t, T)$. It becomes therefore useful to switch to another pricing measure corresponding to an other valuation unit by using the numeraire-change technique (1.44).

For valuation of interest rate claims $\Phi(.)$ the zero-coupon bond $P(t, T)$ has shown to be an appropriate numeraire because of the property $P(T, T) = 1$.
The associated measure is called the *Forward Risk-Neutral Measure*. In this case the change-of-numeraire technique (1.44) can be applied using $P(T, T) = 1$ and $D(t, t) = 1$. This yields a valuation in units of bonds $P(t, T)$ compared to the former valuation in terms of money units with $D(t, T)^6$:

$$\frac{V_t}{\cancel{D(t,t)}}\frac{\cancel{D(t,t)}}{P(t,T)} = \frac{V_t}{P(t,T)} = \mathbb{E}^{RN}\left[\Phi_T \frac{D(t,T)\cancel{P(T,T)}}{\cancel{D(t,t)}P(t,T)}\Big|\mathcal{F}_t\right] =: \mathbb{E}^T[\Phi_t|\mathcal{F}_t]$$

This in turn means a change of the probability measure in the sense of Girsanov's Theorem 1.3.

Theorem 2.2. *Under the* T-Forward Risk-Neutral Measure *the value of a contingent claim $\Phi(.)$ is given by:*

$$V_t = P(t, T)\mathbb{E}^T[\Phi_T|\mathcal{F}_t] \tag{2.28}$$

The Brownian Motion arising from that measure-change:

$$dW_t^T = dW_t^{RN} - \sigma_P(t, T)dt \tag{2.29}$$

is termed T-Forward Brownian Motion.

---

[6]On first view this might be irritating compared with the general numeraire-change (1.44), since we use the discount factor $D(t, T)$ as numeraire multiplicatively here.

Proof. In analogue to our Example 1.1 we have to apply the Product Rule 1.2 to see the relation between the discount factor and bond units:

$$d[\frac{D(t,T)}{P(t,T)}] = D(t,T)d\frac{1}{P(t,T)} + \frac{1}{P(t,T)}dD(t,T) + d\frac{1}{P(t,T)}dD(t,T)$$

and make use of the assumed lognormal process for $P(t,T)$ (2.27).

Furthermore for each $\omega \in \Omega$, the discount factor (2.25) solves the ODE:

$$dD(t,T) = -r(t,\omega)D(t,T)dt$$

A direct calculation using Ito's Lemma reveals, that if we postulate the resulting dynamics $d[\frac{D(t,T)}{P(t,T)}]$ to be driftless in terms of bond units, we necessarily have to set the T-Brownian Motion dW_t^T as we did above.

$\square$

Summing up

Under the introduced forward measure we can now summarize and compare the treated quantities consistently in a model environment.

⋄ At first we would like to point out the difference between the discount factor and the bond price. If the short rate is deterministic $r = r(t)$, we will have $D(t,T) = P(t,T)$. If we assume it to be stochastic, the relationship will be given trough formula (2.26).

⋄ For bond valuation we now have two equivalent expressions one in terms of risk-neutral expectation and the other in terms of forward rates[7].
From that we obtain the risk-neutral forward rate and furthermore

---

[7]This marks the origin for the HJM-Model discussed in Chapter 3

the **Risk-Neutral Yield Curve** (2.12):

$$\left.\begin{array}{l} P(t,T) = \mathbb{E}^{RN}[D(t,T)|\mathcal{F}_t] \\[3mm] P(t,T) = \exp\left[-\int\limits_t^T f(t,\tau)d\tau\right] \end{array}\right\} \Rightarrow$$

$$\boxed{f^{RN}(t,T) \overset{2.10}{=} -\frac{\partial \log \mathbb{E}^{RN}[D(t,T)|\mathcal{F}_t]}{\partial T}} \tag{2.30}$$

◇ Under the forward measure we have the important relationship between short rate and forward rate (for a proof see [9, p.23]) given by:

$$\boxed{\mathbb{E}^T[r_T|\mathcal{F}_t] = f(t,T)} \tag{2.31}$$

◇ The forward measure expresses the fact that a forward rate $F(t; T_1, T_2)$ is the market's arbitrage-free price of the associated future term rate $L(T_1, T_2)$, since it can be constructed as a portfolio of zero-coupon bonds available at time t (see Figure 2.1 and Remark 2.2). This implies for the forward rate to be a martingale under the forward measure.

$$\mathbb{E}^{T_2}[F(T, T_1, T_2)|\mathcal{F}_t] = F(t, T_1, T_2) \quad \forall \quad t \le T \le T_1$$

In particular we have the forward rate as the expected forward-risk neutral value of its corresponding future term rate:

$$\mathbb{E}^{T_2}[L(T_1, T_2)|\mathcal{F}_t] = F(t, T_1, T_2) \tag{2.32}$$

Attaining Consistency between Market and Model

Finally we come back to our announcement from the beginning of Section 2.2 to explain the model valuation of interest rate *securities* (2.14 - 2.16) in terms of forward rate. This is nothing else than a consistent valuation in terms of bond units respectively in terms of the particular forward measure.

To derive those formulas under stochastic interest rates we have implicitly applied the following procedure.

Remark 2.3. *Let there be an arbitrary cash flow from an interest rate security*

$$\mathrm{CF}(.) : \mathbb{R} \rightarrow (0, \infty)$$

depending linearly[8] *on a future term rate $L(T_1, T_2)$ as in formulas (2.14 - 2.16). The term rate as a function on the now stochastic bond price (2.26) becomes itself* a random variable.

The discounted expected risk-neutral value with the stochastic discount factor (2.25) in terms of money units would be given by:

$$V_t = \mathbb{E}^{RN}\big[D(t,T_1)\mathrm{CF}[L(T_1,T_2)]\big|\mathcal{F}_t\big]$$

When changing to a valuation in bond units under the forward measure with Theorem 2.2 the expectation resolves into the associated forward term rate:

$$
\begin{aligned}
V_t &= P(t,T_2)\mathbb{E}^{T_2}\big[\mathrm{CF}[L(T_1,T_2)]\big|\mathcal{F}_t\big] \\
&\overset{Linearity}{=} P(t,T_2)\mathrm{CF}\big[\mathbb{E}^{T_2}[L(T_1,T_2)]\big|\mathcal{F}_t\big] \\
&\overset{2.28}{=} P(t,T_2)\mathrm{CF}\big[F(t,T_1,T_2)\big]
\end{aligned}
$$

---

[8]Each of the introduced securities is a linear function of term rates - in contrast to derivative payoffs

Market Practice and Black's Model

Under the forward measure (2.28) we are now also able to explain the market practice for valuation of plain interest rate derivatives. Black's Model is the analogue to the Black-Scholes Model in the equity world and is widely used in the market to price standard interest rate derivatives such as caps and swaptions. It is based on the *assumption* that a forward term rate $F(t; T_1, T_2)$ evolves with lognormal statistics under the associated forward measure T_2.

$$\boxed{dF(t; T_1, T_2) = \sigma F(t; T_1, T_2)dW_t^{T_2}}\tag{2.33}$$

As we can explicitly solve this SDE (1.21) and obtain the explicit distribution of $F(T, T_1, T_2)$ at $T \geq t$ we also are able to calculate an analytical solution of a derivative payoff using the valuation formula under the forward measure (Theorem 2.2).

We do this exemplarily here for a cap and a payer swaption.

$$V_t^{\text{Cap}} = \mathbb{E}^{RN}[\Phi_{\text{Cap}}|\mathcal{F}_t]$$

$$= \sum_{i=1}^{N} \mathbb{E}^{RN}\left[D(t, T_i)\Delta_i \big[L(T_{i-1}, T_i) - R\big]_+ \Big|\mathcal{F}_t\right]$$

$$\stackrel{2.28}{=} \sum_{i=1}^{N} P(t, T_i)\Delta_i \mathbb{E}^{T_i}\left[\big[F(T_i; T_{i-1}, T_i) - R\big]_+ \Big|\mathcal{F}_t\right]\tag{2.34}$$

We can analytically resolve the expectation in combination with the forward rate dynamics (2.33), for which we refer the reader to Appendix A.2.

For swaptions a similar formula is used, although a main difficulty occurs, since the payoff cannot be additively decomposed as in the case of caps above. Anyway when defining the *Annuity Numeraire* (corresponding to the *Annuity Measure*) using the change-of-numeraire technique (1.44):

$$A_{0,N} := \sum_{i=1}^{N} P(t, T_i) \Delta_i$$

We arrive at the valuation formula:

$$
\begin{aligned}
V_t^{\text{Swaption}} &= \mathbb{E}^{RN}[D(t, T_0) \Phi_{\text{Swaption}} | \mathcal{F}_t] \\
&= \mathbb{E}^{RN}\left[D(t, T_0) \Big[\sum_{i=1}^{N} P(T_0, T_i) \Delta_i [F(t, T_{i-1}, T_i) - R] \Big]_+ \Big| \mathcal{F}_t \right] \\
&= \sum_{i=1}^{N} P(t, T_i) \Delta_i \mathbb{E}^{A_{0,N}}\left[[S_{0,N}(T_0) - R]_+ | \mathcal{F}_t \right]
\end{aligned}
$$

with $S_{0,N}$ the Par Swap Rate (2.17).

A closed formula is also available, when assuming lognormal statistics for the swap rate under that specific swap measure - see [9, pp.20].

3. Models of the Yield Curve

> *In the end, a financial theory is accepted not because it is confirmed by conventional empirical tests, but because researchers persuade one another that the theory is correct and relevant.*
>
> Fischer Black

As we have outlined in the previous chapter there are two main modeling approaches which were developed independently but have shown to be strongly intertwined. The first one is only discussed for completeness, since it is only of secondary relevance for the subsequent chapters.

⋄ *Short Rate Models* define the short rate (2.6) to follow a Markovian Ito Process (1.20) and determine the bond price and the yield curve (2.12) via Theorem 2.1.
Through that the initial (market) yield curve $T \mapsto f^M(0,T)$ as well as its time dynamics is *endogenously* determined via the stochastic bond price (2.26) which in turn depends on the chosen drift and volatility coefficient of the assumed short rate dynamics.

⋄ *Whole Yield Curve Models* model the yield curve directly by assuming stochastic dynamics for the forward rates $f(t,T)$. This approach has shown to be the most general one and is based on the work of D. Heath, R. Jarrow and A. Morton (HJM) . Since the process dynamics are developed just under the restriction of an arbitrage-free market, it provides a general *model class* where many of the above short rate models have shown to be instances of. The yield curve model which has become standard in market practice as it combines the HJM Approach with Black's Model is the *Libor Market Model.*

In this context we would like to refer to [12] for a detailed historical review of yield curve models.

3.1. A Summary of Short Rate Models

Historically interest rate dynamics were first described by so called short-rate models. We mention here the Vasicek Model, the Hull-White Model or the Cox-Ingersoll-Ross (CIR) Model, where for a detailed analysis of those the reader is referred to [9, pp.50]. In addition for a detailed treatment of the Hull-White Model we recommend [9, pp.63]. We summarize here the main common properties of these models.

⬦ The short rate (2.6) gets modeled by a mean-reverting Markovian SDE in the sense of (1.24) which furthermore provides an analytical solution.

Model	Dynamics of the short rate
Vasicek	$dr_t = [\theta - ar_t]dt + \sigma_r dW$
Hull-White	$dr_t = [\theta(t) - ar_t]dt + \sigma_r dW$
CIR	$dr_t = a[\theta(t) - r_t]dt + \sigma_r \sqrt{r}dW$

$$(3.1)$$

⬦ Using Theorem 2.1 and the analytical solvable SDE the model bond price (2.26) as the risk-neutral conditional expectation of the discount factor (2.25) can be calculated analytically and usually assumes the following form:

$$P(t,T) = P(t,T;r_t) = A(t,T)\exp[-B(t,T)r_t] \qquad (3.2)$$

The terms $A(t,T)$ and $B(t,T)$ depend on the particular parameters of the chosen process dynamics of r_t. The exact expressions for $A(t,T)$ and $B(t,T)$ for each model can be found in the above given references. If r_t is normally distributed (as e.g. in the Vasicek and Hull-White Model) and $P(t,T)$ depends exponentially on r_t, $P(t,T)$ will become itself a lognormal distributed random variable.

⋄ As we have the model bond price we can easily derive all other interest rate quantities from Section 2.1. We can see that for instance yields (2.3) are affine functions of the short rate:

$$Y(t,T) = -\frac{\log P(t,T)}{T-t} = \log A(t,T) + B(t,T)r_t$$

We showed the connection of the risk-neutral model bond price and the yield curve in (2.30). Here the risk-neutral dynamics of forward rates $f(t,T)$ are implicitly determined in terms of the assumed dynamics for r_t and with them the dynamics of the whole yield curve (2.12):

$$f^{RN}(t,T) = -\frac{\partial \log P(t,T)}{\partial T} = -\frac{\partial \log A(t,T)}{\partial T} - \frac{\partial B(t,T)}{\partial T}r_t$$

In conclusion: Although we model only one point (namely r_t) on the yield curve, this is sufficient as it determines the numeraire $D(t,T)$ and we can recover bond prices via (2.26) and from those the entire yield curve.

⋄ Modeling the short rate as a stochastic process also implicitly determines the forward rate volatility $\sigma_f(t,T)$. Applying Ito's Lemma to $df(t,T)$ yields the following relationship:

$$\sigma_f(t,T) = -\frac{\partial B(t,T)}{\partial T}\sigma_r$$

Plugging this expression into Black's Formula (A.3), we are able to determine analytically the risk-neutral value for plain derivative products in terms of the short-rate model parameters.

⋄ From those observations also results the main drawback:
The initially observed market yield curve $f^M(0,T)$ is implicitly prescribed by the short-rate model dynamics instead of being an exogenous model input. Although this particular disadvantage can be overcome - for instance in the Hull-White Model, we are still left with an implicit prescription of the *time evolution* of the yield curve through (2.30).

◇ To change to the forward measure in the sense of Theorem 2.2 it is useful to know the resulting bond volatility. Appying Ito's Lemma to $dP(t,T)$ we can see that the general bond volatility coefficient from (2.27) becomes expressed more specifically in terms of the short-rate volatility:

$$\sigma_P(t,T) = -B(t,T)\sigma_r \tag{3.3}$$

◇ Through the assumption of Markovian short rate dynamics, a PDE pricing approach can be applied using the Feyman-Kac Theorem 1.4 in combination with the bond valuation formula (2.26). In the case of the Vasicek Model the bond price would therefore solve[1]:

$$P_t + [\theta - ar]P_r + \frac{1}{2}\sigma_r^2 P_{rr} = rP$$

At the end of the current section we sum up in four steps the procedure which is characteristic for a short-rate model:

◇ Assume the short rate r_t to be some realistic Markovian Process.

◇ The short rate determines the money unit numeraire i.e. the stochastic discount factor $D(t,T)$ through (2.25).

◇ The risk-neutral conditional expectation of $D(t,T)$ corresponds to the risk-neutral bond price via Theorem 2.1.

◇ The yield curve and its dynamics is then determined via the connection between bond prices and forward rates (2.30).

---

[1]Under the use of subscripts for partial derivatives

3.2. The Heath-Jarrow-Morton Framework

The essential idea of D. Heath, R. Jarrow and A. Morton[2] is to model the risk-neutral time evolution of the *entire* yield curve (2.12) exogenously.

This is achieved by assuming general stochastically perturbed dynamics for each single forward rate $f(t, T) = f(t, T; \omega)$ directly:

$$df(t, T) = \alpha(t, T)dt + \sigma_f(t, T)dW_t \quad \forall \quad T \geq t \tag{3.4}$$

From those general dynamics the risk-neutral specification $df^{RN}(t, T)$ can be derived through the forward rate's connection to the numeraire - see (2.30).

In particular this determines the general introduced drift term $\alpha(t, T)$ under such a *Bond Risk-Neutral Measure*, i.e. under absence of arbitrage between bonds with different maturities. It turns out that specifying the forward rate volatilities $\{\sigma_f(t, T)\}_{T \geq t}$ in combination with the initial market yield curve $f^M(0, T)$ as exogenous model input is sufficient to determine the entire risk-neutral curve dynamics $df^{RN}(t, T)$.

This approach has shown to be a general risk-neutral model framework, where other interest rate models have been recovered as special instances - in particular all short-rate models.

Beneath its generality one of its main advantages is that in the general setting the initially observed market yield-curve $f^M(0, T)$ serves as a direct exogenous model input.

---

[2]Bond Pricing and the Term Structure of Interest Rates: A new methodology for Contingent Claim Valuation, *Econometrica*, 60(1):77-105, January 1992

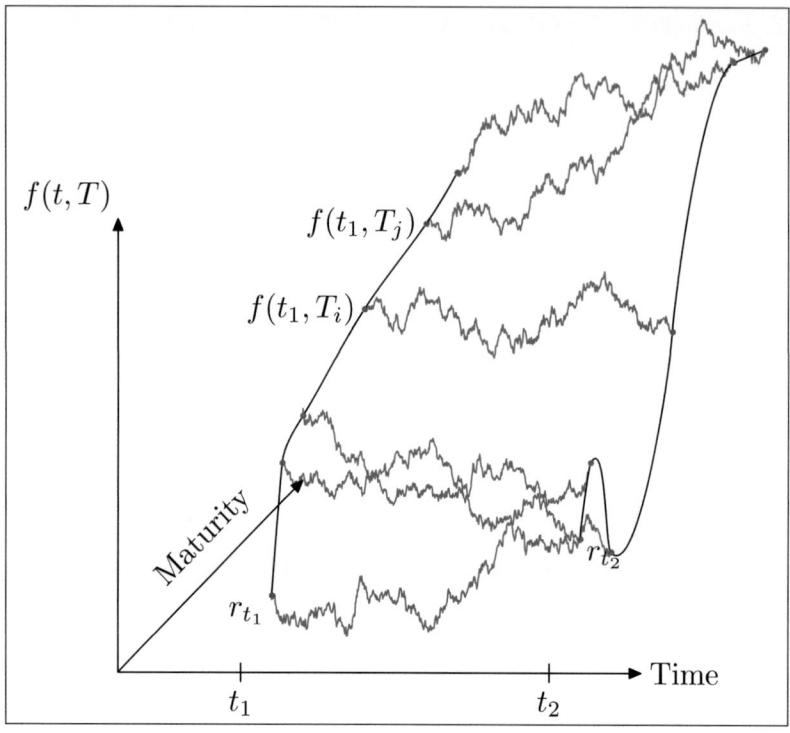

Figure 3.1.: Evolution of the Yield Curve driven by Stochastic Forward Rates

A pure No-Arbitrage Argument

Starting with the above general dynamics (3.4) it was shown, what the drift term $\alpha(t, T)$ has to satisfy with respect to the risk-neutral measure under which discounted bonds with different maturities are martingales.

This was achieved by using the martingale property of the discounted bond (2.27) in conjunction with the expression of the bond price in terms of forward rates (2.30).

Theorem 3.1 (One-Factor HJM). *If there does not exist arbitrage between bonds $P(t,T)$ with different maturities T, each corresponding forward rate $f(t,T)$ with maturity T evolves separately according to:*

$$df^{RN}(t,T) = \left[\sigma_f(t,T)\int_t^T \sigma_f(t,\tau)d\tau\right]dt + \sigma_f(t,T)dW_t^{RN} \quad (3.5)$$

where the drift term $\alpha(.,.) : [0,T] \times [0,T^*] \to \mathbb{R}$ from (3.4) is fully specified by the forward rate volatility $\sigma_f(.,.) : [0,T] \times [0,T^*] \to \mathbb{R}$.

Proof. Integration (for simplicity from 0 to t) of the above SDE (3.4) yields under the risk-neutral measure[3]:

$$f(t,T;\omega) = f(0,T) + \int_0^t \alpha(s,T)ds + \int_0^t \sigma_f(s,T)dW_s^{RN}(\omega)$$

For the short rate we have:

$$r_t(\omega) = f(0,t) + \int_0^t \alpha(s,t)ds + \int_0^t \sigma_f(s,t)dW_s^{RN}(\omega) \quad (3.6)$$

Therefore the stochastic discount factor (2.25) is given by:

$$D(0,t) \overset{2.25}{=} \exp\left[-\int_0^t f(0,\tau)d\tau - \int_0^t\int_0^\tau \alpha(s,\tau)ds d\tau \right.$$
$$\left. -\int_0^t\int_0^\tau \sigma_f(s,\tau)dW_s^{RN}d\tau\right]$$

The arbitrage free - i.e. discounted driftless - bond price dynamics are taken from the dynamics (2.27):

$$d\big[D(0,t)P(t,T)\big] = \sigma_P(t,T)dW^{RN}$$

---

[3]If the context is clear we just write $f(t,T)$ instead of $f^{RN}(t,T)$

The bond price's expression in terms of forward rates by formula (2.11) yields on the other hand:

$$P(t,T) = \exp\left[-\int_t^T f(0,\tau)d\tau - \int_t^T \int_0^\tau \alpha(s,\tau)dsd\tau\right.$$
$$\left.-\int_t^T \underbrace{\int_0^\tau \sigma_f(s,\tau)\,dW_s^{RN}}_{:=-\sigma_P(t,\tau)}d\tau\right]$$

After interchanging integration order[4] and establishing a relationship between forward and bond volatilites to obtain consistency with the chosen bond-price dynamics

$$\sigma_P(t,T) := -\int_t^T \sigma_f(t,\tau)d\tau \tag{3.7}$$

The discounted bond price in terms of forward rates writes:

$$D(0,t)P(t,T) =$$
$$\exp\left[\underbrace{-\int_0^T f(0,\tau)d\tau - \int_0^T \int_s^T \alpha(s,\tau)d\tau ds + \int_0^T \sigma_P(s,T)dW_s^{RN}}_{:=X}\right]$$

Applying Ito to $d\left[e^X\right] = e^X dX + \frac{1}{2}e^X dX dX$ we arrive at the following dynamics:

$$d[D(0,t)P(t,T)] = D(0,t)P(t,T)\left[\left[\frac{1}{2}\sigma_P(t,T)^2 - \int_t^T \alpha(t,\tau)d\tau\right]dt\right.$$
$$\left.+ \sigma_P(t,T)dW^{RN}\right]$$

As we are working under the risk-neutral measure under which the discounted bond is a martingale, the drift term has to vanish which necessarily implies:

$$\frac{1}{2}\sigma_P(t,T)^2 = \int_t^T \alpha(t,\tau)d\tau$$

Solving this equation for $\alpha(t,T)$ means differentiating with respect to T. Therefore in the risk-neutral setting, the arbitrarily introduced drift term $\alpha(t,T)$ of the general assumed dynamics in (3.4) now becomes completely determined by the bond or forward rate volatility coefficient:

$$\alpha(t,T) = \sigma_P(t,T)\frac{\partial \sigma_P(t,T)}{\partial T} = \sigma_f(t,T)\int_t^T \sigma_f(t,\tau)d\tau$$

Summarizing the proof: The martingale dynamics of the discounted bond price imply specific dynamics for the associated forward rate through formula (2.30).

$\square$

Remark 3.1. *Forward Rates have a non-zero drift under the risk neutral measure - equivalently to a valuation in terms of money units. In contrast to the forward measure (2.32) they are not martingales.*

Remark 3.2. *Under the risk-neutral forward rate dynamics (3.5) the dynamics for the short rate process r_t is not Markovian. With the drift term $\bar{\alpha}(t,T)$ as in (3.5) and the short-rate as in (3.6) we have:*

$$dr_t = df(t,T)\Big|_{T=t} + \frac{\partial f(t,T)}{\partial T}\Big|_{T=t} dt$$

$$= \left[\bar{\alpha}(t,T) + \frac{\partial f(0,T)}{\partial T} + \int_0^t \frac{\partial \bar{\alpha}(s,T)}{\partial T}ds + \int_0^t \frac{\partial \sigma_f(s,t)}{\partial T}dW_s\right]_{T=t} dt$$

$$+ \sigma_f(t,t)dW_t^{RN} \tag{3.8}$$

This yields a Path Dependence *due to the stochastic integral inside the drift coefficient.*

---

[4] For interchanging integration order when stochastic integrals are involved, there is also a stochastic Fubini Theorem which justifies this procedure

Remark 3.3 (Multi-Factor HJM). *We would like to point to a multi-factor version of the HJM forward rate dynamics:*

$$df(t,T) = \sum_{k=1}^{n} \sigma_k(t,T) \int_t^T \sigma_k(t,s)ds + \sum_{k=1}^{n} \sigma_k(t,T)dW_k^{RN} \quad (3.9)$$

where $W_1, \ldots, W_n$ are independent Brownian Motions under the risk-neutral measure and with $\sigma_f(.,.) : [0,T] \times [0,T^*] \to \mathbb{R}^n$.

Discussion

We sum up the general HJM-Approach as a model having essential theoretical advantages but practical drawbacks.

⬦ The result provides a general **Framework** of risk-neutral exogenous yield curve modeling corresponding to the risk-neutral equity model in the Black-Scholes environment.

⬦ The yield curve dynamics (3.5) are completely determined once the structure of the forward rate volatility $\sigma_f(t,T)$ is specified and furthermore the initial market yield curve data $f^M(0,t)$ are included.

⬦ By *specifying the volatility structure* in this general *class* of yield curve models, one recovers short-rate models or the Libor Market Model (Section 3.3) as instances. With the specification

$$\sigma_f(t,T) := \bar{\sigma}e^{-a(T-t)} \quad (3.10)$$

we arrive at the Hull-White short-rate dynamics. We will come back to this point in the next chapter and for a derivation of the Libor Market Model see the following section.

To work with the dynamics (3.5) with a general volatility structure $\sigma_f(t,T)$ has essential drawbacks from an implementation perspective.

⋄ The short rate dynamics (3.8) is not Markovian in general. It
  is therefore not possible to apply a PDE valuation as a conjunction between the Bond-Pricing Theorem 2.1 and the Feynman-Kac
  Theorem 1.4. The pricing has usually to be done via Monte-Carlo
  Simulation.

⋄ The general HJM dynamics does *not* allow the yield curve evolution to be represented by a *finite dimensional* state space, since the
  yield curve (2.12) consists of an uncountable number of forward
  rates $f(t, T)$. When discretizing the time scale as in the Libor
  Market Model (see Section 3.3) this results in high-dimensional
  dynamics.

⋄ To value a security depending on the yield curve evolution the
  curve dynamics gets modeled through a high number of forward
  rates $\{f(t, T)\}_{T \geq t}$, where each of those has to be simulated separately.
  This increases computational time with complexity of the product
  - meaning the number of forward rates a product depends on.

⋄ To make the general HJM consistent with market practice, which
  is dominated by Black's Formula and the assumption on the forward rate dynamics (2.33), we would assume the statistics for the
  forward rate to be lognormal distributed. One straightforward
  approach to do this is to set $\sigma(t, T) := \bar{\sigma} f(t, T)$, with $\bar{\sigma}$ constant,
  which yields:

$$df(t, T) = \bar{\sigma}^2 f(t, T) \int_t^T f(t, u) du + \bar{\sigma} f(t, T) dw$$

Since the drift is non-random, the solution of that SDE will not
be lognormal. Even worse the drift is quadratic in $f(t, T)$, which
leads to the instability of the solution, the so-called *Lognormal
Explosion*. This property is discussed in [13, pp. 248].

3.3. The Libor Market Model

The aim of Market Models - especially the *Libor Market Model* - introduced 1997 by A. Brace, D. Gatarek and M. Musiela[5] (therefore also sometimes called *BGM Model*) was to *synthesize* Market Practice, which had been dominated by Black's Model (2.33), with an *Arbitrage Free Yield Curve Model*, whose fundamental approach had to be consistent within the HJM Framework.

Due to the upcoming of more complex interest-rate derivatives, whose payoffs depend on more than one forward rate, the problem with Black's Formula arose how to relate more than one forward rate under one single pricing measure.

Resolving this issue the Libor Market Model has shown to be the closed representative of the HJM Model-Framework. This approach directly models observable market term rates in contrast to idealized forward rates and is consistent with market practice. For a reference see [14, pp.577] or [11, pp. 230].

The Model Environment

Although not introduced this way the Libor Market Model (LMM) has shown to be a discretized HJM with lognormal dynamics, which can be derived - comparable to the case of short rate models - directly from the HJM setting (Theorem 3.1) by a specification of the forward rate volatility $\sigma_f(t, T)$. Furthermore the dynamics are determined not on a continuous time scale but on a discretized tenor structure $\mathcal{T} = \{T_0, \ldots, T_N\}$. Therefore the model dynamics cannot be formulated with respect to the continuously compounded cash-account numeraire $D(t, T)$ as in the HJM. An appropriate numeraire has to be chosen for the discretized time scale. It is represented by a constant reinvestment strategy at a tenor point T_i into a bond with maturity T_{i+1}.

$$\text{SL}(t) := \prod_{i=1}^{N} \frac{1}{P(T_{i-1}, T_i)} = \prod_{i=1}^{N} \left[1 + L(T_{i-1}, T_i)\Delta_i\right] \tag{3.11}$$

and is termed the *Spot Libor Numeraire* (SL) [11, p.233] or *Rolling Commercial Deposit (CD) [14, p.578]*.

---

[5] The Market Model of Interest Rate Dynamics, *Mathematical Finance*, 7(2):127-147, January 1997

A direct Derivation from the general HJM Dynamics

We give now a derivation of the LMM-Forward-Rate-Dynamics in the one-factor case starting from the gerenal HJM-Dynamics (3.5) in the risk-neutral measure. Afterwards we explain what measure-change has to be done to get to a valuation on a discretized time scale in terms of the above introduced numeraire $SL(t)$.
Start with a Forward Term Rate (2.9) - here denoted as $L_k(t) := F(t, T_k, T_{k+1})$ with fixing point T_k and maturity T_{k+1}.

$$1 + L_k(t)\Delta_k \overset{2.11}{=} \frac{P(t, T_k)}{P(t, T_{k+1})} = \exp\left[\underbrace{\int_{T_k}^{T_{k+1}} f(t, \tau)d\tau}_{:=X}\right] = e^X$$

The differential version is given by an application of Ito's Lemma 1.1:

$$d[1 + L_k(t)\Delta_k] = \Delta_k dL_k(t) = e^X dX + \frac{1}{2}e^X dX dX$$

$$= e^X\left[d\left[\int_{T_k}^{T_{k+1}} f(t, \tau)d\tau\right] + \frac{1}{2}d\left[\int_{T_k}^{T_{k+1}} f(t, \tau)d\tau\right]d\left[\int_{T_k}^{T_{k+1}} f(t, \tau)d\tau\right]\right]$$

We have started from the risk-neutral measure where we know the explicit dynamics of $f(t, T)$ through (3.5). Furthermore we are able to change the order of the time and maturity differentials which yields:

$$\Delta_k dL_k(t) \overset{3.5}{=} e^X\left[\int_{T_k}^{T_{k+1}} \left[\sigma_f(t, \tau)\int_t^\tau \sigma_f(t, u)du\right]d\tau\right.$$

$$+ \frac{1}{2}\int_{T_k}^{T_{k+1}} \sigma_f(t, \tau)d\tau \int_{T_k}^{T_{k+1}} \sigma_f(t, \tau)d\tau\right]dt + e^X\left[\int_{T_k}^{T_{k+1}} \sigma_f(t, \tau)d\tau\right]dW_t^{RN}$$

To obtain consistency with Black's assumption (2.33) we would like to obtain a lognormal structure for the forward term rate, so we choose the forward rate volatility to be:

$$\boxed{\sigma_f(t,\tau) \equiv \frac{L_k(t)\sigma_k(t)}{1 + L_k(t)\Delta_k} \quad \forall \quad \tau \in [T_k, T_{k+1})}$$ (3.12)

With that specification the integral preceding the Brownian Motion term resolves into:

$$\int_{T_k}^{T_{k+1}} \sigma_f(t,\tau)d\tau = \frac{L_k(t)\sigma_k(t)}{1 + L_k(t)\Delta_k}\Delta_k$$

As an intermediate result we now rewrite the lognormal dynamics (compare to 1.21) with the new volatility coefficient. In addition we undo the above replacement $e^X = 1 + L_k(t)\Delta_k$:

$$dL_k(t) = \underbrace{\frac{1 + L_k(t)\Delta_k}{\Delta_k}\bar{\mu}\,dt}_{:=\mu^{LMM}} + L_k(t)\sigma_k dW_t^{RN}$$

We are left with the drift term μ^{LMM} whose calculation in terms of the above volatility structure is done now. Firstly we make use of the volatility specification (3.12) to obtain:

$$\mu^{LMM} = \frac{1 + L_k(t)\Delta_k}{\Delta_k}\left[\int_{T_k}^{T_{k+1}}\left[\sigma_f(t,\tau)\int_t^\tau \sigma_f(t,u)du\right]d\tau\right.$$

$$\left. + \frac{1}{2}\int_{T_k}^{T_{k+1}}\sigma_f(t,\tau)d\tau\int_{T_k}^{T_{k+1}}\sigma_f(t,\tau)d\tau\right]$$

$$= \frac{L_k(t)\sigma_k(t)}{\Delta_k}\left[\int_{T_k}^{T_{k+1}}\left[\int_t^\tau \sigma_f(t,u)du\right]d\tau + \frac{1}{2}\frac{L_k(t)\sigma_k(t)}{1 + L_k(t)\Delta_k}\Delta_k^2\right]$$

Since our volatility coefficient (3.12) is piecewise constant, we can discretize the inner integral on the discrete tenor structure $[T_{m(t)}\ldots T_k]$ where we further define:

$$m(t) := \operatorname{argmin}_{i\in\{0,\ldots,N\}}(T_i \geq t)$$ (3.13)

So we have:

$$\mu^{LMM} = \frac{L_k(t)\sigma_k(t)}{\Delta_k} \left[\int_{T_k}^{T_{k+1}} \left[\int_t^{T_{m(t)}} \sigma_f(t,u)du + \sum_{j=m(t)}^{k-1} \Delta_j \sigma_f(t,T_j) \right. \right.$$

$$\left. + (\tau - T_k)\sigma_f(t,\tau) \right] d\tau + \frac{1}{2} \frac{L_k(t)\sigma_k(t)}{1 + L_k(t)\Delta_k} \Delta_k^2 \Bigg]$$

$$= \frac{L_k(t)\sigma_k(t)}{\Delta_k} \left[\int_{T_k}^{T_{k+1}} \left[\int_t^{T_{m(t)}} \sigma_f(t,u)du + \sum_{j=m(t)}^{k-1} \Delta_j \frac{L_j(t)\sigma_j(t)}{1 + L_j(t)\Delta_j} \right] d\tau \right.$$

$$\left. \underbrace{- \int_{T_k}^{T_{k+1}} (\tau - T_k)\sigma_f(t,\tau)d\tau}_{(1)} + \frac{1}{2} \frac{L_k(t)\sigma_k(t)}{1 + L_k(t)\Delta_k} \Delta_k^2 \right]$$

Now (1) resolves into:

$$(1) = \int_{T_k}^{T_{k+1}} (\tau - T_k) \frac{L_k(t)\sigma_k(t)}{1 + L_k(t)\Delta_k} d\tau$$

$$= \frac{1}{2}\tau^2 - T_k\tau \Big|_{T_k}^{T_{k+1}} \frac{L_k(t)\sigma_k(t)}{1 + L_k(t)\Delta_k} = \frac{1}{2} \frac{L_k(t)\sigma_k(t)}{1 + L_k(t)\Delta_k} \Delta_k^2$$

Therefore we arrive at:

$$\mu^{LMM} = \frac{L_k(t)\sigma_k(t)}{\Delta_k} \left[\Delta_k \int_t^{T_{m(t)}} \sigma_f(t,\tau)d\tau \right.$$

$$\left. + \sum_{j=m(t)}^{k-1} \int_{T_k}^{T_{k+1}} \Delta_j \frac{L_j(t)\sigma_j(t)}{1 + L_j(t)\Delta_j} d\tau + \Delta_k^2 \frac{L_k(t)\sigma_k(t)}{1 + L_k(t)\Delta_k} \right]$$

$$= L_k(t)\sigma_k(t) \left[\int_t^{T_{m(t)}} \sigma_f(t,\tau)d\tau \right.$$

$$\left. + \sum_{j=m(t)}^{k-1} \Delta_j \frac{L_j(t)\sigma_j(t)}{1 + L_j(t)\Delta_j} + \Delta_k \frac{L_k(t)\sigma_k(t)}{1 + L_k(t)\Delta_k} \right]$$

$$= L_k(t)\sigma_k(t) \sum_{j=m(t)}^{k} \Delta_j \frac{L_j(t)\sigma_j(t)}{1 + L_j(t)\Delta_j} + L_k(t)\sigma_k(t) \int_t^{T_{m(t)}} \sigma_f(t,\tau)d\tau$$

We now rewrite the entire dynamics with the above calculated drift term μ^{LMM} still with respect to the risk-neutral measure.

$$
dL_k = \left[L_k(t)\sigma_k(t) \left[\sum_{j=m(t)}^{k} \Delta_j \frac{L_j(t)\sigma_j(t)}{1 + L_j(t)\Delta_j} + \int_t^{m(t)+1} \sigma_f(t,\tau)d\tau \right] \right] dt
$$
$$
+ \quad L_k(t)\sigma_k dW^{RN}
$$

The corresponding measure of the chosen rolling forward numeraire in the sense of (1.44) is termed the *Rolling Forward Risk-Neutral Measure* [14, p.578] or *Spot Libor Measure (SL)*. It is characterized by the property that each forward term rate has driftless dynamics on one particular tenor interval. In our setting we specify the measure change (in the sense of (1.30) from the risk-neutral to this Spot Libor Measure by:

$$
dW_t^{SL} := \left[\int_t^{T_{m(t)}} \sigma_f(t,\tau)d\tau \right] dt + dW_t^{RN} \tag{3.14}
$$

Here the drift term which describes the difference between the two measures can be seen as a *residual* term from the continuous-time version of the HJM, which does not fit into the discretized tenor-structure of the LMM.

The final term-rate dynamics in the Libor Market Model with (3.13) are now given by:

$$
\boxed{ dL_k = \left[L_k(t)\sigma_k(t) \sum_{j=m(t)}^{k} \Delta_j \frac{L_j(t)\sigma_j(t)}{1 + L_j(t)\Delta_j} \right] dt + L_k(t)\sigma_k dW_t^{SL} }
$$

$$
\tag{3.15}
$$

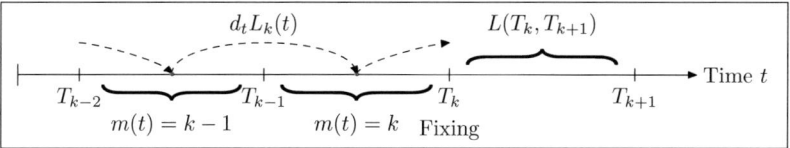

Figure 3.2.: Drift-Evolution in the Libor Market Model

Figure 3.2 shows how a particular Libor Rate $L_k(t)$ with fixing date T_k gets modeled in the Libor Market Model. Especially we see that *every* Libor Rate L_k has driftless dynamics on the discrete tenor interval directly before its particular fixing date: $T_k - T_{k-1}$.

PDE Valuation in the LMM

Now we are exactly at the crossroads, which we mentioned in the preface of the book. With (3.15) we found a time-discrete version of the general HJM and beyond with (3.12) we introduced lognormal behavior consistently with Black's Model. But are we able to do a PDE Valuation?

The answer is theoretically *YES*. But we have to take a look on the dimension of the underlying process. In the general setting we have from the process dynamics(3.15) $k - m(t)$ state variables. Since this would yield $k - m(t)$ spatial dimensions for the corresponding Feynman-Kac PDE, we have to negate the question from a practical - i.e. numerical - viewpoint, which will be made clearer later on.

However we would like to emphasize that PDE Valuation is surely possible to value Caplets. For a Caplet on a Term Rate $L_i(t)$ and Payoff (2.21) at time T_i the dynamics of the Libor Market Model (3.15) resolves into the forward rate dynamics of Black's Model (2.33).

Under its associated forward measure, the value of the Caplet is therefore given by:

$$V_t(x) = P(t, T_i)\mathbb{E}^{T_i}\left[\left[L_i(T_i) - K\right]_+ \Big| L_i(t) = x\right]$$

As assumption (2.33) implies Markovian Dynamics, we are able to apply the Feynman-Kac Theorem (1.4) to find that we can resolve the expectation by PDE methods. In particular we find:

$$\frac{\partial V}{\partial t} + \frac{1}{2}\sigma_{L_i}^2 \frac{\partial^2 V}{\partial x^2} = 0$$

This helps us in the case when we assume complex specifications of the volatility structure σ_{L_i} (see also Section 6.2), as we are not able to resolve the above expectation analytically.

How this can be used effectively in calibration procedures is discussed in ([15]).

Conclusion

When we consider the just derived dynamics for a particular Libor Rate L_k (3.15) we see that the drift term depends on outcomes of other Libor Rates L_j ($j = m(t) \ldots k$) on the discretized tenor structure. Therefore this exogenous yield-curve model is Markovian but due to its dependence on other rates only in terms of a high-dimensional state space.

This now motivates the question if there is a *low* Markovian system capturing the yield-curve dynamics exogenously where still being consistent within the HJM-Framework (3.5).

4. Markovian Representations of the Yield Curve

> *When you try out a new yield curve model, you're implicitly saying something like "Let's pretend people in markets care only about the level of future short-term interest rates, and they expect them to be be distributed normally". As you say that to yourself, if you're honest, your heart sinks.*

> Emanuel Derman

As shown in Chapter 3 in absence of arbitrage between bonds with different maturities, the risk-neutral dynamics (3.5) of every single instantaneous forward rate $f^{RN}(t, T)$ is fully specified by the forward rate volatility $\sigma_f(t, T)$ and the initial market yield curve $f^M(0, T)$. In the general case (3.5) the state space of the entire yield curve $T \mapsto f(t, T)$ is infinite dimensional and often high dimensional (see the LMM Dynamics (3.15)). Furthermore the general forward rate dynamics imply path-dependent dynamics for the short rate (see Remark 3.2).

On the other hand it has been recovered in several works - see e.g. [1] and [16] that choosing a specific volatility structure $\sigma_f(t, T)$ yields an exogenous model of the yield curve with Markovian dynamics. We present this procedure here following [1] where a so-called *separable volatility structure* has been proposed.

In the first section we derive the resulting dynamics directly from (3.5) with a volatility parameter separated into two factors which yields a two-dimensional Markovian SDE System.

This is followed by an analysis of the implication for the short-rate dynamics which also shows to be Markovian.

We then derive the formula of the risk-neutral bond price in terms of the Markovian state variables. Due to the underlying Markovian System we are able to give the valuation PDE for general payoffs using the Feynman-Kac Theorem (1.4).

Afterwards we analyze the specific case of specifying the parameters in the SDE system to be constant, which gives some nice insight of how this model is related to the previously discussed short-rate models. In the last section of this chapter we give a short outlook what the model dynamics become in a more general multi-factor volatility setting originally also derived in [1].

4.1. Separable Volatility: The Cheyette Model

We already remarked in the discussion of the HJM at the end of Section 3.2 that with a specification of the forward rate volatility $\sigma_f(t,T)$ one obtains the Markovian Hull-White Model dynamics.

This approach can be generalized by assuming the volatility to be separable into time and maturity dependent factors.
With that one arrives at a subclass of yield-curve models, where the risk-neutral HJM dynamics (3.5) can be represented through a higher dimensional Markovian SDE system. We now analyze the two-factor case.

Theorem 4.1 (Markovian Yield Curve). *Let the forward-rate volatility* $\sigma_f(t,T)$ *in (3.5) be separable into a Markovian process* $H : [0,T] \times \Omega \to \mathbb{R}$ *and a maturity dependent deterministic function* $g : [0,T^*] \to \mathbb{R} \setminus \{0\}$:

$$\sigma_f(t,T) := H_t g(T) \tag{4.1}$$

Then the yield curve (2.12) at a future time $t > 0$ *- which evolves according to the risk-neutral forward rate dynamics (3.5) - is Markovian and becomes determined by two state variables:*

$$f(t,T) = f(0,T) + \beta(t,T)\mathbf{X_t} + \beta(t,T) \int_t^T \beta(t,\tau)d\tau \mathbf{Y_t} \tag{4.2}$$

where $\beta(t,T)$ *is defined as:*

$$\beta(t,T) := \exp\left[-\int_t^T \kappa(\tau)d\tau \right] \tag{4.3}$$

The state variables X_t *and* Y_t *are Markovian Processes. Their coupled dynamics are given by:*

$$dX_t = \left[Y_t - \kappa(t)X_t\right]dt + \eta_t dW_t^{RN} \tag{4.4}$$

$$dY_t = \left[\eta_t^2 - 2\kappa(t)Y_t\right]dt \tag{4.5}$$

with initial values $X_0 = 0$ *(w.p.1) and* $Y_0 = 0$*. The parameters* η *and* κ *defined as*[1]:

$$\eta_t := H_t g(t) \quad and \quad \kappa(t) := -\frac{g'(t)}{g(t)}. \tag{4.6}$$

Remark 4.1. *In terms of the new defined parameters* κ *and* η *the forward rate volatility structure of* $\sigma_f(t,T)$ *has a specific form as we have from (4.1) with* $\sigma_f(t,T) := H_t g(t)\frac{g(T)}{g(t)}$:

$$\sigma_f(t,T) = \eta_t \beta(t,T) \tag{4.7}$$

---

[1] $g'(t)$ denotes the first derivative with respect to t

The above model setup is usually termed *Cheyette Model* in the newer literature due to the work of O. Cheyette [1], although this relationship was also independently recovered in other works (e.g. [16]).

Proof. Time-Integration from time 0 to $t \leq T$ of the HJM SDE (3.5) and replacing $\sigma_f(t, T)$ with the above factorization yields:

$$f(t, T) = f(0, T) + \int_0^t \left[H_s g(T) \int_s^T H_s g(\tau) d\tau \right] ds + \int_0^t H_s g(T) dW_s$$

The important point in the derivation is that we can extract the time-dependent term out of the inner maturity integral and the maturity-dependent term out of the Ito-Integral:

$$f(t, T) = f(0, T) + g(T) \left[\int_0^t \left[H_s^2 \int_s^T g(\tau) d\tau \right] ds + \int_0^t H_s dW_s \right]$$

Furthermore we split the inner integral with upper bound T into parts and factor out $\frac{1}{g(t)}$. We are now able to define two stochastic factors which only depend on time until time-point t:

$$f(t, T) = f(0, T) + \frac{g(T)}{g(t)} \left[g(t) \underbrace{\left(\int_0^t \left[H_s^2 \int_s^t g(\tau) d\tau \right] ds + \int_0^t H_s dW_s \right)}_{:=X_t} \right.$$

$$\left. + g(t)^{-1} \underbrace{g^2(t) \int_0^t H_s^2 ds \int_t^T g(\tau) d\tau}_{:=Y_t} \right]$$

$$= f(0, T) + \frac{g(T)}{g(t)} \left[X_t + Y_t g(t)^{-1} \int_t^T g(\tau) d\tau \right]$$

Using the fact from the definition of $\kappa(t)$ in (4.6):

$$\frac{g(T)}{g(t)} = \exp \left[\int_t^T \frac{g'(s)}{g(s)} ds \right] = \exp \left[\int_t^T -\kappa(s) ds \right]$$

we obtain (4.2).

The dynamics of the new defined stochastic process X_t are given through the following SDE:

$$
\begin{aligned}
dX_t &= d\left[g(t)\left(\int_0^t \left[H_s^2 \int_s^t g(\tau)d\tau \right]ds + \int_0^t H_s dW_s \right) \right] \\
&= \left[\frac{g'(t)}{g(t)}X_t + g(t)\int_0^t H_s^2 g(t)ds \right]dt + H_t g(t)dW_t \\
&= \left[\frac{g'(t)}{g(t)}X_t + Y_t \right]dt + H_t g(t)dW_t
\end{aligned}
$$

Since not explicitly driven by a Brownian Motion Increment the second state variable Y_t will only be stochastic, if H_t is chosen to be a stochastic process.

$$
dY_t = \left[H_t^2 g^2(t) + 2\frac{g'(t)}{g(t)}Y_t \right]dt
$$

As a special case if H_t is chosen to depend only on time $H_t = H(t)$ or constant, the dynamics of Y_t will result into an ordinary differential equation.

Furthermore we can now see, why the process H_t has to obey the Markov Property itself, because otherwise the whole SDE System (4.4) would loose its Markovian structure. Restricting H_t to become a function of particular current system state (X_t, Y_t) we are able to define new parameters as in (4.6). With those our volatility structure for $\sigma_f(t, T)$ becomes restricted as in (4.7):

$$
\eta_t := \eta(t, X_t, Y_t) := H(t, X_t, Y_t)g(t) \quad \Rightarrow \quad H(t, X_t, Y_t) = \frac{\eta(t)}{g(t)}
$$

where g(t) solves an ODE with initial value $c \neq 0$:

$$
\kappa(t) := -\frac{g'(t)}{g(t)} \quad \Rightarrow \quad g(T) = c\exp\left[-\int_t^T \kappa(\tau)d\tau \right] \quad \text{with} \quad g(t) = c
$$

The desired Markovian Dynamics for X_t and Y_t as well as the resulting volatility structure from Remark (4.7) are thereby proven.   □

We now sum up how the process H_t and in turn the parameter η_t can be specified:

⋄ Choosing H_t to be constant or time dependent we have a deterministic ODE for Y_t and therefore only one (stochastic) Markov dimension in X_t. We refer to Section 4.4 for a further treatment.

⋄ To hold the Markovian structure H_t can be chosen to depend on the current system state (X_t, Y_t). In this case we do have two Markov dimensions. We treat this more general case in the following sections.

⋄ Beneath its dependence on (X_t, Y_t) H_t can also be chosen depend on or to be driven by additional Markovian Processes, which then would yield three or more Markov dimensions. How this is used in this current model setup is explained in Section 6.2.

We have presented here the risk-neutral dynamics of $f(t, T)$ in integrated form (4.2). As those determine the short-rate and the bond price, which we will explain in the following, we do not really need the explicit time dynamics $df(t, T)$ as in (3.5). For completeness they are still provided at the end of Section 4.2 (see Proposition 4.4).

Implications for the short-rate dynamics

Unlike the general HJM model - see Remark 3.2 - the short rate in the above separable volatility setting (4.1) *is* Markovian. In this sense the model specification is similar to the previously discussed Markovian short rate model.

Proposition 4.1. *In the separable volatility setting the evolution of the short-rate (2.13) is characterized through Markovian dynamics:*

$$\boxed{r_t = f(0,t) + X_t} \tag{4.8}$$

Written in differential form we have:

$$dr_t = \underbrace{\left.\frac{\partial f(s,T)}{\partial T}\right|_{(s,T)=(0,t)}}_{:=\partial_T f(0,t)} + dX_t$$

$$= \left[\partial_T f(0,t) + Y_t + \kappa[f(0,t) - r_t]\right]dt + \eta_t dW_t \tag{4.9}$$

Proof. From its definition in (2.6) we have:

$$r_t := \left. f(t,T)\right|_{T=t}$$

So we simply replace $T = t$ in (4.2) and obtain (4.8). □

Interpretation of Parameters and State Variables

⋄ The parameter $\kappa(t)$ has the interpretation of being the mean-reversion speed, where the stochastic parameter η_t denotes the volatility of the short-rate.

Reviewing the properties of mean-reverting processes from Section 1.5, we can nicely observe the implications of mean-reversion on the forward rate volatility structure in formula (4.7):

With an high mean-reversion, the process tends to converge to its long term mean, which is equivalent of saying the forward rate volatility (4.7) tends to decrease fast over time. The reverse effect would apply for low mean-reversion parameters.

⋄ The state variable X_t has a direct interpretation of *linking* each point $f(t, T)$ on the current yield curve to its associated risk-neutral short-rate value r_T at a future time T.

⋄ The state variable Y_t is driven only implicitly stochastic (due to its dependence of η_t) and can be interpreted of being the *integrated variance* of the particular forward rate with maturity t: $f(\cdot, t)$. This can be seen right from its definition in the proof of Theorem 4.1:

$$Y_t = \int_0^t H_\tau^2 g^2(t) d\tau = \int_0^t \sigma_f^2(\tau, t) d\tau$$

4.2. The Analytical Bond Price

We are able to give an analytic expression of the bond price in terms of the two introduced state variables using formula $(2.11)^2$.

Proposition 4.2. *Given $X_t = x$ and $Y_t = y$ at time t the price of a zero-coupon bond becomes expressed through:*

$$P(t, T; x, y) = \frac{P(0, T)}{P(0, t)} \exp \left[-G(t, T)x - \frac{1}{2}G(t, T)^2 y \right] \qquad (4.10)$$

where we define with (4.3):

$$G(t, T) := \int_t^T \beta(t, \tau) d\tau \qquad (4.11)$$

$\frac{P(0,T)}{P(0,t)}$ *denotes the* Market Forward Bond Price *for borrowing at time t with maturity T which is obtained from market discount curve (2.2) $P^M(0, T)$ - see Figure 2.1.*

Proof. We start with formula (2.11) and write $f(t, T)$ in terms of our new state variables (see previous proof):

$$P(t, T) = \exp \left[-\int_t^T f(t, \tau) d\tau \right]$$

$$= \exp \left[-\int_t^T f(0, \tau) d\tau \right]$$

$$\exp \left[-\int_t^T X_t \frac{g(\tau)}{g(t)} d\tau - \int_t^T \left[Y_t \frac{g(\tau)}{g^2(t)} \int_t^\tau g(s) ds \right] d\tau \right]$$

---

[2]Since we work under the risk-neutral measure, we will obtain the same result by solving the bond valuation formula (2.26).

[2]We have here from the product rule: $2 \int_t^T v'(\tau)v(\tau) d\tau = v(\tau)^2 \big|_t^T$ with $v(\tau) := \int_t^\tau \frac{g(s)}{g(t)} ds$

Using the following the definition of $\kappa(t)$ from (4.6) and $\beta(t,T)$ from (4.3) we arrive at:

$$P(t,T) = \frac{P(0,T)}{P(0,t)} \exp\left[-X_t \underbrace{\int_t^T \beta(t,\tau)d\tau}_{:=G(t,T)} -Y_t \int_t^T \frac{g(\tau)}{g(t)} \left[\int_t^T \frac{g(s)}{g(t)} ds \right] d\tau \right]$$

To cope with the integral behind the second state variable Y_t we make use of the product rule of integration[3] and obtain the desired formula again with the above replacement with $\kappa(t)$:

$$P(t,T) = \frac{P(0,T)}{P(0,t)} \exp\left[-G(t,T)X_t - \frac{1}{2}Y_t \left[\underbrace{\int_t^T \beta(t,\tau)d\tau}_{=G(t,T)} \right]^2 \right]$$

$$= \frac{P(0,T)}{P(0,t)} \exp\left[-G(t,T)X_t - \frac{1}{2}G(t,T)^2 Y_t \right]$$

$\square$

Remark 4.2. *The bond price was calculated from formula (2.11). On the other hand we also have from Theorem 2.26 that:*

$$P(t,T;x,y) = \mathbb{E}^{RN}[D(t,T)|X_t = x, Y_t = y]$$

Remark 4.3. *An important feature of formula (4.10) is that through $G(t,T)$ the bond price depends only on the mean-reversion parameter κ. The volatility is in some way implicitly described by the state variable y - see Remark 4.1. This becomes essentially important in view of realistic modeling - see Section 6.3.*

In many recent works (for instance [17]) concerning the implementation the parameter κ is chosen to be constant, so we can give a more direct expression for $G(t,T)$:

Remark 4.4. *If κ is chosen to be constant, then direct integration yields:*

$$G(t,T) = \frac{1}{\kappa}\left[1 - \exp(-\kappa(T-t))\right]$$

Corresponding to the bond-price formulas within the main short-rate models (3.2) we can also express the bond-price in terms of the short-rate using relationship (4.8).

Remark 4.5. *The bond price in terms of the short rate writes:*

$$P(t,T) = A(t,T) \exp\left[-G(t,T)r_t \right]$$

with

$$A(t,T) := \frac{P(0,T)}{P(0,t)} \exp\left[G(t,T)f(0,t) - \frac{1}{2}G(t,T)^2 Y_t \right]$$

The bond volatility of the resulting dynamics $dP(t,T)$ is derived in the same manner as in (4.3). We would like to give a different proof of the result which comes directly from the volatility specification in the HJM model.

Proposition 4.3. *The bond volatility is given by:*

$$\sigma_P(t,T) = -G(t,T)\eta_t$$

Proof. From the HJM we have an explicit relationship between the bond volatilities and the forward volatilities (3.7):

$$\sigma_P(t,T) = -\int_t^T \sigma_f(t,\tau)d\tau$$

Using the separable volatility structure

$$\sigma_P(t,T) = -\int_t^T H_t g(\tau)d\tau = -\underbrace{H_t g(t)}_{=\eta_t}\int_t^T \frac{g(\tau)}{g(t)}d\tau$$

and again the definition of κ in combination with (4.3) we obtain the desired result. □

To close the circle with Section 4.1 as we now have the explicit bond price formula (4.10), we come back to the risk-neutral HJM dynmics (3.5) and show what they become explicitly in terms of the dynamics of the state variables X_t and Y_t.

This is now derived circuitously from the model bond price (4.10) in the sense of a short-rate model (see Section 3.1).

Proposition 4.4. *The dynamics of the forward rate $f(t,T)$ is given in terms of the state variables (X_t, Y_t) by:*

$$df(t,T) = \partial_T f(0,t)dt + \beta(t,T)dX_t + \beta(t,T)G(t,T)dY_t$$

Proof. Take the logarithm of the bond price (4.10):

$$-\log P(t,T) = -\log P(0,T) + \log P(0,t) + G(t,T)X_t + \frac{1}{2}G(t,T)^2 Y_t$$

from which we calculate the forward rate:

$$f(t,T) = \frac{\partial \log P(0,T)}{\partial T} + \frac{\partial \log P(0,t)}{\partial T} + \beta(t,T)X_t + \beta(t,T)G(t,T)Y_t$$

$$= f(0,T) - f(0,t) + \beta(t,T)X_t + \beta(t,T)G(t,T)Y_t$$

Using Ito's Lemma 1.1 we obtain the dynamic version:

$$df(t,T) = \frac{\partial f(t,T)}{\partial t}dt + \frac{\partial f(t,T)}{\partial x}dX_t + \frac{\partial f(t,T)}{\partial y}dY_t + \underbrace{\frac{\partial^2 f(t,T)}{\partial x^2}dX_t dX_t}_{=0}$$

$$= \frac{\partial f(t,T)}{\partial t}dt + \beta(t,T)dX_t + \beta(t,T)G(t,T)dY_t$$

□

4.3. The Valuation PDE

Using our new Markovian system (4.4) we are now able to derive a two-dimensional valuation PDE for any derivative payoff Φ depending on an interest rate quantity as a straightforward application of the Feynman-Kac Theorem 1.4 in combination the dynamics in (4.4).

Theorem 4.2 (PDE Valuation). *Under the separable volatility structure (4.1) the expected value $V(t, x, y)$ of any forward rate dependent contingent claim $\Phi(.,.) : \mathbb{R}^2 \to \mathbb{R}$*

$$V := V(t, x, y) = \mathbb{E}^{RN}\left[\exp\left[-\int_t^T r_\tau d\tau\right]\Phi(X_T, Y_T)\Big| X_t = x, Y_t = y\right]$$
(4.12)

solves the following parabolic PDE:

$$0 = \frac{\partial V}{\partial t} + \left[y - \kappa x\right]\frac{\partial V}{\partial x} + \frac{1}{2}\eta^2(t, x, y)\frac{\partial^2 V}{\partial x^2} - \underbrace{\left[f(0, t) + x\right]}_{=r_t} V$$

$$+ \left[\eta^2(t, x, y) - 2\kappa y\right]\frac{\partial V}{\partial y}$$
(4.13)

Note that the diffusion term in the second space dimension y is not existent due to the missing Brownian Motion increment in the dynamics of the second state variable Y_t.

To connect the main formulas of this chapter developed so far consider the following proposition which connects the bond price (4.10) to the above PDE.

Proposition 4.5. *The value of the bond price $P(t, T)$ given through formula (4.10) solves the above parabolic PDE (4.13) with final-time condition $\Phi(x_T, y_T) \equiv 1$.*

Proof. We proof the result using constant κ. The bond price pays one unit at its maturity T so we have

$$V(T, x, y) = \mathbb{E}^{RN}\left[\mathbf{1}\Big|X_T = x, Y_T = y\right] = 1$$

Starting from formula (4.10) we first calculate the needed partial derivatives where to maintain overview, we make use of the following shorthand notations:
$P_x := \frac{\partial P}{\partial x}$, $G := G(t, T)$, $\hat{P} := \frac{P(0,T)}{P(0,t)}$, $A := -G(t,T)x - \frac{1}{2}G^2(t,T)y$.
Using these we have:

$$P_x = -\hat{P}Ge^A, \quad P_{xx} = -\hat{P}G^2e^A, \quad P_y = -\frac{1}{2}\hat{P}G^2e^A$$

$$P_t = \frac{\partial\hat{P}}{\partial t}e^A + \hat{P}\Big[\frac{\partial}{\partial t}\Big[-Gx - \frac{1}{2}G^2y\Big]e^A\Big]$$

$$= \Big(\frac{\partial\hat{P}}{\partial t} + \hat{P}\Big[-x\frac{\partial G}{\partial t} - yG\frac{\partial G}{\partial t}\Big]\Big)e^A$$

With the following relationship

$$G = \frac{1}{\kappa}\big[1 - \exp(-\kappa(T-t))\big] \Rightarrow \frac{\partial G}{\partial t} = -\exp[-\kappa(T-t)] = \kappa G - 1$$

We get for the time derivative after differentiating the first summand w.r.t. t:

$$P_t = -\hat{P}\underbrace{\frac{1}{P(0,t)}\frac{\partial P(0,t)}{\partial t}}_{=-f(0,t)}e^A + \hat{P}\Big[-x(\kappa G - 1) - yG(\kappa G - 1)\Big]e^A$$

We now put all derivatives together in formula (4.13) and after canceling $\hat{P}e^A$:

$$f(0,t) + \big[x - x\kappa G + yG - y\kappa G^2\big] + (\kappa x - y)G + \frac{1}{2}\eta^2 G^2$$

$$-\frac{1}{2}(\eta^2 - 2\kappa y)G^2 - f(0,t) - x \quad = \quad 0$$

$\square$

4.4. The Case of Constant Parameters

To embed the model setup into the context of short-rate models from
Section 3.1, we like to discuss the case of constant parameters η and κ.
In this setting - essentially since η itself is not stochastic - we can simplify our system (4.4) by solving the resulting deterministic ordinary
differential equation for the second state variable $Y_t{}^4$. This allows us
to consider a one-dimensional SDE from which we can derive a one-
dimensional PDE and even analytical solutions for standard derivative
payoffs. Furthermore we will see that this particular setting corresponds
exactly to the Hull-White Model - see Table 3.1.

Recovering the Hull-White-Model Dynamics

As we already noted in the Discussion of the HJM at the end of Chapter 3 many short rate models can be recovered from the general HJM
setting through a specification of the volatility coefficient. To embed
our analysis of separable volatility into that environment we derive the
complete Hull-White Dynamics (see [9, pp.63]) within the setting of the
current chapter.

Proposition 4.6. *The system (4.4) with specified constant parameters*
$\eta_t \equiv \eta$ *and* $\kappa(t) \equiv \kappa$ *leads to the Hull-White Model Dynamics:*

$$dr_t = \left[\theta(t) - \kappa r_t\right]dt + \sigma dW_t^{RN}$$

where $\theta(t)$ *is given by*

$$\theta(t) = \partial_T f(0,t) + \kappa f(0,t) + \frac{\sigma^2}{2\kappa}\left[1 - e^{-2\kappa t}\right]$$

---

[4]This will also apply for the case, if $\eta = \eta(t)$ is chosen to be time-dependent.

Proof. In the sense of (4.7) we have here - compare this to what has already been mentioned at the end of Section 3.2 (see 3.10):

$$\sigma_f(t,T) = \eta e^{-\kappa(T-t)}$$

In terms of the two functions from Theorem 4.1 we could also specify $g(T) := \sigma e^{-\kappa T}$ and $H_t := e^{\kappa t}$. Rewriting the dynamics (4.4) yields:

$$dX_t = \big[y - \kappa X_t\big]dt + \eta dW_t$$
$$dY_t = \big[\eta^2 - 2\kappa Y_t\big]dt$$

We see that the dynamics of Y satisfy a inhomogeneous linear ordinary differential equation, which can be solved analytically on the interval $[0,t]$ using the method of Variation of Constants[5]:

$$Y(t) = e^{-2\kappa t}\big[c + \frac{\eta^2}{2\kappa}[e^{2\kappa t} - 1]\big] \quad \text{with} \quad Y(0) = c$$

The Hull-White-Dynamics are a special solution when choosing the initial value to be zero. We make use of the Markovian short rate dynamics (4.9) to arrive in combination with the above result at:

$$dr_t = \partial_T f(0,t)dt + dX_t$$
$$= \partial_T f(0,t) + \big[\frac{\eta^2}{2\kappa}(1 - e^{-2at}) - \kappa X_t\big]dt + \eta dW_t$$

and finally using from (4.8) that $X_t = r_t - f(0,t)$ we obtain:

$$dr_t = \big[\partial_T f(0,t) + \kappa f(0,t) + \frac{\eta^2}{2\kappa}(1 - e^{-2at}) - \kappa r_t\big]dt + \eta dW_t^{RN}$$

$$\square$$

Furthermore we see, that this special solution of $Y(t)$ corresponds exactly to the Variance of the Ornstein-Uhlenbeck Process (1.26). In other words for the Hull-White Dynamics the integrated forward rate variance of the forward rate (see Remark 4.1) becomes incorporated explicitly into the drift-term.

---

[4]See e.g. O. Forster, Analysis 2, pp.115

Analytical Solution for Derivative Payoffs

We are able to derive an analytical formula for European derivatives when assuming constant parameters as we can solve the resulting one-dimensional SDE analytically. The following formula was firstly developed by F. Jamshidian[6].

$$dX_t = \left[Y_t - \kappa X_t\right]dt + \eta dW_t^{RN}$$

$$= \left[\frac{\sigma^2}{2\kappa}[1 - e^{-2\kappa t}] - \kappa X_t\right]dt + \eta dW_t^{RN} \tag{4.14}$$

We have already solved this kind of SDE in Proposition 1.3 so we simply replace in (1.24) $b := \frac{\sigma^2}{2\kappa}[1 - e^{-2\kappa t}]$, $a := \kappa$ and $\sigma := \eta$. With $X_0 = 0$ we then have for the solution:

$$X_t \sim \mathcal{N}\left[\frac{\eta^2}{2\kappa^2}\left[1 - e^{-\kappa t}\right]^2, \frac{\eta^2}{2\kappa}\left[1 - e^{-2\kappa t}\right]\right]$$

As we have explained in Section 2.4 for evaluating contingent claims, the forward measure (2.28) is very useful. We therefore switch to this pricing measure in the sense of Theorem 2.2 which implies changes of the drift coefficient:

$$dX_t = \left[Y_t - \kappa X_t - \eta\sigma_P(t,T)\right]dt + \eta dW_t^T$$

We do not have to calculate the exact distribution of X_t with respect to this measure.

Instead we are using the following argumentation: Since the short rate via the stochastic process X_t is gaussian distributed, the bond price is lognormal distributed as it depends exponentially on X_t (4.10). Furthermore we know that under the forward measure from Theorem 2.28 we have the relationship for a today's value of a bond-value at a future time T with maturity $T_1 > T$:

$$\mathbb{E}^T[P(T,T_1)|\mathcal{F}_0] = \frac{P(0,T_1)}{P(0,T)}$$

where $P(0,T_1)$ and $P(0,T)$ are known quantities directly out of the market.

---

[6]An Exact Bond Option Formula, *Journal of Finance*, 44:205-209, January 1989

As we already have a formula for the bond volatility in Proposition 4.3 we have now for the variance of the bond price:

$$\text{Var}_{P(t,T)} = G(t,T)^2 \text{Var}(X_t) = \frac{\eta^2}{2\kappa^3}\left[1 - e^{-2\kappa(T-t)}\right]^2\left[1 - e^{-2\kappa t}\right]$$

Using Black's Formula (A.3) for quantities under their Forward Measure we are now able to evaluate exemplarily the value of a zero-coupon bond call (2.18) with maturity T_1 on a Bond P_t with maturity $T_2 \geq T_1$ at valuation time $t = 0$:

$$\begin{aligned}
V_0 &= P(0,T_1)\mathbb{E}^{T_1}\left[\left[P_T - K\right]_+\right] \\
&= P(0,T_2)N(d_1) - KP(0,T_1)N(d_2)
\end{aligned} \tag{4.15}$$

with

$$d_1 = \frac{\log\frac{P(0,T_2)}{P(0,T_1)} - \log K + \frac{1}{2}\sigma_P^2}{\sigma_P}$$

$$d_2 = d_1 - \sigma_P$$

$$\sigma_P := \sqrt{\text{Var}_{P(T,T_1)}}$$

The One-Dimensional Valuation PDE

Applying the Feynman-Kac Theorem 1.4 to the above calculated now one-dimensional dynamics of the yield curve (4.14) in terms of X_t, we obtain the PDE Dynamics with one space dimension for the risk-neutral expectations (4.12) in the sense of Theorem 4.2:

$$0 = \frac{\partial V}{\partial t} + \left[\frac{\sigma^2}{2\kappa}[1 - e^{-2\kappa t}] - \kappa x\right]\frac{\partial V}{\partial x} + \frac{1}{2}\eta^2\frac{\partial^2 V}{\partial x^2} - [f(0,t) + x]V \tag{4.16}$$

4.5. Multi-Factor Volatility

Also not further treated in this thesis we would like to draw the reader's attention to the fact that the two-factor Markovian representation of the yield curve in Theorem 4.1 can be extended to a more general multi-factor setting (see [1] or [2]). We introduce this case only for showing that also more state variables in the Markovian setting are still more advantageous than to track explicitly a huge family of forward rates respectively term rates as in the Libor Market Model (3.15).

Theorem 4.3. *Starting from a multifactor HJM (3.3), assume that*

$$\sigma_f^i(t, T) = \sum_{k=1}^{n} g_k(T) H_k^i(t, \omega) \quad \forall i = 1 \ldots n \tag{4.17}$$

with $g : [0, \infty) \to \mathbb{R}^n \setminus \{0\}$ and $H : [0, \infty) \times \Omega \to \mathbb{R}^{n \times n}$. Then we have the Markovian System[7] with $i = 1 \ldots n$:

$$dX_i = \Big[\sum_{j=1}^{n} Y_i^j - \kappa_i(t) X_i \Big] dt + \sum_{j=1}^{n} \eta_i^j dW_j^{RN}$$

$$dY_j^i = \Big[\sum_{k=1}^{n} \eta_k^i \eta_k^j - (\kappa_i(t) + \kappa_j(t)) Y_j^i \Big] dt \quad j = 1 \ldots i$$

with

$$\eta^i(t, \omega) = \sum_{k=1}^{n} g_k(t) H_k^i(t, \omega)$$

and

$$\frac{g(T)}{g(t)} = \Big[e^{-\int_t^T \kappa_1(\tau) d\tau}, \ldots, e^{-\int_t^T \kappa_n(\tau) d\tau} \Big]$$

---

[7] We write row (=upper) and column (=lower) indices

Remark 4.6. *In the above n-dimensional volatility separation structure and due to the symmetry property of $dY_j^i = dY_i^j$ we have:*

$\diamond$ *n state variables $X^i(t, \omega)$*

$\diamond$ *$\frac{n(n+1)}{2}$ state variables $Y_j^i(t, \omega)$*

The question is how many factors are necessary to obtain a flexible model which is capable of reproducing different market phenomena. This becomes essentially important when the model gets calibrated to market data, where it has to match diverse shapes of implied volatility surfaces in the derivatives market - for further details we refer to Sections 6.4 and 6.5.

In [1, pp.7] it was empirically observed that with a two or three factor model (respectively 5 and 9 state variables) one is able to match more complex shapes of volatility surfaces.

In comparison to the dynamics of the Libor Market Model (3.15), when e.g. a four factor model will be used to price a 30 year structure with quarterly compounding - i.e. depending on four term rates per annum, we would have to simulate 120 state variables, whereas in the current setting we would have to track only 14. For further details see also [2].

Part II.

Numerics and Implementation

5. Numerical Solution

An approximate answer to the right problem is worth a good deal more than an exact answer to an approximate problem.

John Tukey

In this chapter we investigate the numerical approximation of the Feynman-Kac PDE (1.34) and in particular the two-dimensional version of the Valuation PDE from Theorem 4.2. We choose the method of *Finite Differences* combined with an operator splitting algorithm known as the **A**lternating **D**irection **I**mplicit Method which was especially developed to efficiently solve finite difference schemes for PDE's with multiple spatial dimensions.

After introducing some important discretization methods for partial derivatives we explain the general procedure and important finite difference schemes for one spatial dimension, which we also put into a probability theoretical context. After a discussion on stability and convergence we explain the general ADI procedure for the original Peaceman-Rachford scheme and afterwards we analyze our used ADI algorithm based on the work of [18]. This is followed by a discussion on analytical and numerical boundary conditions and an investigation of the improvement of accuracy by working on non-equidistant grids.

5.1. Discretization of Differential Opterators

Before introducing finite difference schemes we would like to show how to approximate partial derivatives of a function $u(t, x)$. An approximation is easily obtained on a discretized equidistant grid consisting of M points with constant grid size

$$h := x_m - x_{m-1} \quad \forall m = 1 \dots M$$

as a linear combination of Taylor expansions in different directions and step sizes with grid corresponding approximations $U_m \approx u(x_m)$.

In general we can say that the (local) order of accuracy increases with the number of used discretization points which are included in the discretization (e.g. fourth order when using 5 grid points). With an increasing number of used grid points we will have more freedom to linear combine additional Taylor expansions to cancel higher order terms. We demonstrate the procedure here exemplarily for three point discretizations.

Three-Point Discretizations of first and second partial derivatives

Using three discrete space points x_{m-1}, x_m, x_{m+1} we obtain an approximation at x_m by:

$$U_{m-1} = U_m - \frac{\partial u}{\partial x}h + \frac{1}{2}\frac{\partial^2 u}{\partial x^2}h^2 + O(h^3)$$

$$U_{m+1} = U_m + \frac{\partial u}{\partial x}h + \frac{1}{2}\frac{\partial^2 u}{\partial x^2}h^2 + O(h^3)$$

Subtraction and division by h yields a second order accurate approximation:

$$\frac{\partial u}{\partial x} = \frac{U_{m+1} - U_{m-1}}{2h} + O(h^2)$$

In turn when adding both Taylor expansions we obtain an equally accurate approximation for the second order derivative:

$$\frac{\partial^2 u}{\partial x^2} = \frac{U_{m+1} - 2U_m + U_{m-1}}{h^2} + O(h^2)$$

We have listed some more discretizations in the following table, which can all be obtained in a similar manner. For further discretizations see also Appendix A.6.

	Operator	U_{i-2}	U_{i-1}	U_i	U_{i+1}	U_{i+2}	Denom	Order
$\frac{\partial u}{\partial x}$	$\mathcal{D}_x^{2-}$	0	-1	1	0	0	h	$O(h)$
	$\mathcal{D}_x^{2+}$	0	0	-1	1	0	h	$O(h)$
	$\mathcal{D}_x^{3-}$	1	-4	3	0	0	$2h$	$O(h^2)$
	$\mathcal{D}_x^{3}$	0	-1	0	1	0	$2h$	$O(h^2)$
	$\mathcal{D}_x^{3+}$	0	0	-3	4	-1	$2h$	$O(h^2)$
	$\mathcal{D}_x^{5}$	1	-8	0	8	-1	$12h$	$O(h^4)$
$\frac{\partial^2 u}{\partial x^2}$	$\mathcal{D}_{xx}^{3}$	0	1	-2	1	0	h^2	$O(h^2)$

$$(5.1)$$

In the following we mainly make use of difference operators involving at most three grid points. The resulting matrices arising when applying the particular combination of difference operators $\mathcal{D}$ at the grid approximations U_m of the entire grid $U = [U_0 \ldots U_M]^T \in \mathbb{R}^{M+1}$ usually then have a tridiagonal structure.

The resulting linear equation systems can be solved efficiently with asymptotic computational complexity $O(N)$ in dependence of the dimension of the system $N := M+1$ by the use of the following algorithm.

An Efficient Direct Solver for Tridiagonal Matrices

For completeness we would like to review a modified Gauss Elimination Algorithm without Pivoting. This algorithm is especially suited for tridiagonal matrices, which are diagonal dominant - in particular which do not have a zero diagonal element. It is known as the *Thomas Algorithm* in the literature (see e.g. [19, pp.23]).

First of all it is known that the general Gauss Elimination Algorithm with Pivoting has asymptotic complexity of $O(N^3)$.

Now given a tridiagonal linear system $AU = b$, in each row of the corresponding matrix $A \in \mathbb{R}^{N \times N}$, there are at most three non-zero elements - the lower diagonal, upper diagonal and diagonal elements. This writes:

$$d_0 U_0 + u_0 U_1 = b_0$$
$$l_m U_{m-1} + d_m U_m + u_m U_{m+1} = b_m \quad (m = 1 \dots M - 1)$$
$$l_M U_{M-1} + d_M U_M = b_M$$

Where for all $m = 0 \dots M$ and $u_M = l_0 = 0$:

$$|d_m| > |u_m| + |l_m|$$

The algorithm can be split up into a *Forward Step* working through the rows from 1 to M followed by a reversely working *Backward Step*.
Firstly the forward step forms an upper-tridiagonal matrix by elimination of the lower-diagonal elements l_m:

$$\bar{d}_m U_m + u_m U_{m+1} = \bar{b}_m \quad (m = 1 \dots M)$$

with the new diagonal and upper diagonal elements:

$$\bar{d}_m = d_m - \frac{l_m}{d_{m-1}} u_{m-1} \quad \text{and} \quad \bar{b}_m = b_m - \frac{l_m}{d_{m-1}} b_{m-1}$$

In the subsequent backward substitution we can solve this recursively starting with U_M:

$$U_M = \frac{\bar{b}_M}{\bar{d}_M} \quad \text{and} \quad U_m = \frac{\bar{b}_m - \bar{u}_m U_{m+1}}{\bar{d}_m} \quad (m = M - 1 \dots 0)$$

For each single row there are exactly 3 Multiplications + 2 Divisions + 3 Subtractions, which is equal to 8 Floating Point Operations. This yields *linear* asymptotic complexity $O(N)$.

5.2. Finite Difference Schemes

The approach of finite differences is the most intuitive one for numerically solving a PDE. The procedure is summarized by building a discrete time and space grid on which the differential operators can be replaced by above introduced difference operators $\mathcal{D}$. The accuracy of this local approximation depends on the truncation error which in turn depends on the chosen discretization.

Applying this procedure transforms the differential equation into a sequence of linear equation systems, whose each single dimension depend on the spatial grid dimension. The number of systems to be solved is overall equal to the number of chosen time steps.

The Feynman-Kac PDE revisited

We would like to investigate the main finite difference schemes for a *specified* Feynman-Kac PDE (1.34), where we now neglect mixed partial derivatives. Although this PDE is valid on the entire $\mathbb{R}^q$, for its numerical solution we now set the spatial region to be a q-dimensional rectangular domain:

$$\Omega := \{(x_1 \ldots x_q) \in \mathbb{R}^q \mid l_i \leq x_i \leq u_i, \quad i = 1 \ldots q\}$$

and further define a time and space region $[T_0, T_1] \times \Omega$, where $[T_0, T_1]$ is a subset of our entire previously defined time interval $[0, T^*]$. The PDE (1.34) from Theorem 1.4 is a spatial q-dimensional homogeneous convection-diffusion PDE with time and space-dependent coefficients, which is solved on the above spatial region backwards in time[1]. With the notations and definitions as in Theorem 1.4 we have:

$$\frac{\partial u}{\partial t} + \mathcal{L}u = 0 \quad \text{with} \quad u(T_1, x) = \Phi(x) \tag{5.2}$$

---

[1]In the literature, finite difference schemes are usually applied to initial value problems, which are solved forward in time. In view of our valuation purpose of a future payoff $\Phi(.)$, we are solving our PDE backwards in time. We could also easily transform our PDE into an initial value problem by transforming from *Time* to *Time To Maturity*: $t \to T - t$

From now on we specify the spatial Differential Operator (1.36) to be of the following form:

$$\mathcal{L} := \sum_{i=1}^{q} D_{x_i} - r(t,x)u \quad \text{with} \tag{5.3}$$

$$D_{x_i} := a_i(t,x)\frac{\partial}{\partial x_i} + b_i(t,x)\frac{\partial^2}{\partial x_i^2}$$

and $x := (x_1, \ldots, x_q) \in \mathbb{R}^q$. For a solution $u(t,x)$, which we define from (1.33) as $u(t,x) := V(t,x)$, we assume:

$$u(t,x) \in \mathcal{C}^1\big([T_0, T_1)\big) \times \mathcal{C}^2(\Omega)$$

We proceed considering some additional *limiting assumptions* for the subsequent treatment concerning the general PDE (5.2).

Convection and Dissipation Behavior

◇ The *Convection* coefficients $a_i(t,x) \in \mathbb{R}$ *transport* (=convect) the initial profile given by $\Phi(x)$ through the spatial domain with *speeds* $a_i(x,t)$. In view of the underlying Markovian process X_t in the sense of Feynman-Kac the convection coefficient corresponds to the drift term of the underlying Markovian process.

◇ The *Diffusion* coefficient $b_i(t,x)$ which we define from (1.34) as $b_i(t,x) := \frac{1}{2}[\sigma^T\sigma]_{i,i}$ (therefore always positive: $b_i(t,x) \geq 0$) is determined by the volatility coefficients of the underlying process. In view of the PDE (5.2) it is responsible for the *Dissipation* - or Damping - behavior of the spatial operator $\mathcal{L}$ with *strengths* $b_i(t,x)$ which smooth out the final-time profile $\Phi(.)$ over time.

◇ Finally the function $r(t,x) \geq 0$ - usually interpreted as the interest rate of the discount factor in (1.33) - determines the *Reaction* coefficient.

Difference Schemes - General Procedure

To solve the above PDE one firstly has to define the time and space grid, which is supposed to be placed on the defined time and space region. Furthermore in addition to the explicitly known final-time condition $\Phi(x)$ (1.35) additional boundary conditions at the domain boundaries have to be specified to obtain a unique numerical solution[2].

We now introduce a uniform grid $[T_0, T_1]_k \times \Omega_h$, defined by the numbers of time-steps N and the respective numbers of spatial steps M_i.

Using those we obtain the according grid-sizes:

$$(\Delta_t; \Delta x_1, \ldots \Delta x_q)$$

by defining:

$$\Delta_t = \frac{T_1 - T_0}{N} \quad \text{and} \quad \Delta x_i = \frac{u_i - l_i}{M_i}$$

On the grid the exact solution $u(x,t)$ becomes approximated by the corresponding grid approximation $U_m^n := U(n; m_1 \ldots m_q)$ with $0 \leq m_i \leq M_i$ and $0 \leq n \leq N$:

$$U_m^n = u(t_n; x_m) + O(\Delta_t^r) + \sum_{i=1}^{q} O(\Delta_{x_i}^s)$$

with $x_m := [x_{m_1}, \ldots, x_{m_q}]^t$, $t_n := T_1 - n\Delta_t$ and $x_{m_i} = l_i + m_i \Delta_{x_i}$.

Here the local orders of accuracy r and s depend mostly on the applied discretization operator. A simple *one time-step*[3] finite difference scheme is now constructed at each discrete time point t_n by replacing the time-partial derivative by a two-point backward discretization $\mathcal{D}_t^{2-}$. This yields a sequence of linear equation systems for the approximations U^n and U^{n-1} which is solved backwards for all time steps t_n.

---

[2]The Feynman-Kac PDE 1.34 is valid on the entire space $\mathbb{R}^n$, so there is now explicit information available concerning boundary behavior on self-chosen space boundary conditions

[3]For Multilevel Schemes involving more than one time step see [20, pp.300]

Definition 5.1. *We define a finite difference scheme for our PDE (5.2) with a backward difference for the time derivative* $\mathcal{D}_t^{2-}$ *as:*

$$FD_{k,h}U^n := \frac{U^n - U^{n-1}}{\Delta_t} + \theta \mathcal{L}_h U^{n-1} + (1-\theta)\mathcal{L}_h U^n = 0 \quad (5.4)$$

with the spatial discretized operator $\mathcal{L}_h$, *which is the sum of the particular spatial* difference *operators* $\bar{D}$. *Those in turn are constructed by suitable discretizations* $\mathcal{D}$ *from Table 5.1.*

$$\mathcal{L}_h := \sum_{i=1}^{q} \bar{D}_{x_i} - r(t_n, x_m) \quad \text{with}$$

$$\bar{D}_{x_i} := a_i(t_n, x_m)\mathcal{D}_x + b_i(t_n, x_m)\mathcal{D}_{xx}$$

We further redefine the time and space step sizes as $k := \Delta_t$ *and* $h_i := \Delta_{x_i}$.

The spatial operator is applied either entirely to the approximation of the already calculated point in time t_n with $\theta = 0$ - which yields an full *Explicit* scheme - or to the one to be calculated t_{n-1} with $\theta = 1$ - which yields a full *Implicit* scheme. Beyond that also *weighted* schemes with $0 < \theta < 1$ are possible.

We now give some specifications of (5.4) in the case of one spatial dimension ($q = 1$) and constant coefficients.

The Explicit Scheme

Following [5, p.129] we use central discretizations for the space derivatives from which we can obtain first order time $O(k)$ and second order space $O(h^2)$ accuracy. Applying $\mathcal{L}_h$ to the already calculated grid point U_m^n and rewriting the system equation (5.4) with $\theta = 0$ at a single grid point (n, m) yields the explicit expression for U_m^{n-1}:

$$U_m^{n-1} = (1 - 2b\nu - rk)U_m^n + b\nu(1+\alpha)U_{m+1}^n + b\nu(1-\alpha)U_{m-1}^n$$

$$(5.5)$$

where we define:

$$\nu := \frac{k}{h^2} \quad \text{and} \quad \alpha := \frac{a}{2b}h$$

We see that each grid point at the next time-step U_m^{n-1} is simply a weighted sum of grid-points from the previous time-step. Therefore the scheme is easily solved iteratively through time. However it suffers from restrictions in view of the choice of k and h, therefore it may not be *unconditional stable*. This is because as soon as the factors become negative, the grid approximation may not satisfy the *Maximum Principle* anymore although the exact solution $u(t, x)$ does:

$$\sup_x |u(t, x)| \leq \sup_x |u(T, x)| \quad \text{if} \quad t < T$$

The Implicit Scheme

With the same notations as above but applying the spatial difference operator to the space points U_m^{n-1} of the time point to be calculated with $\theta = 1$ yields:

$$\boxed{(1 + 2b\nu - rk)U_m^{n-1} - b\nu(1 + \alpha)U_{m+1}^{n-1} + b\nu(\alpha - 1)U_{m-1}^{n-1} = U_m^n}$$

$$(5.6)$$

Writing this equation for all spatial grid points yields a linear equation system with tridiagonal structure at each point in time t_n. As explained in the previous section this can be solved - in combination with appropriate assumptions on the boundary behavior - efficiently with linear complexity by the Thomas algorithm.

The Weighted Scheme

The above one-step schemes have the property that they are only first order accurate in time due to the two point time discretization. This accuracy can be increased at intermediate time points t_θ by applying $\mathcal{L}_h$ weighted to U^n and U^{n-1} with a weighting parameter θ. We do not write out the resulting equation explicitly since it is just a θ-weighted equation of the two above introduced equations.

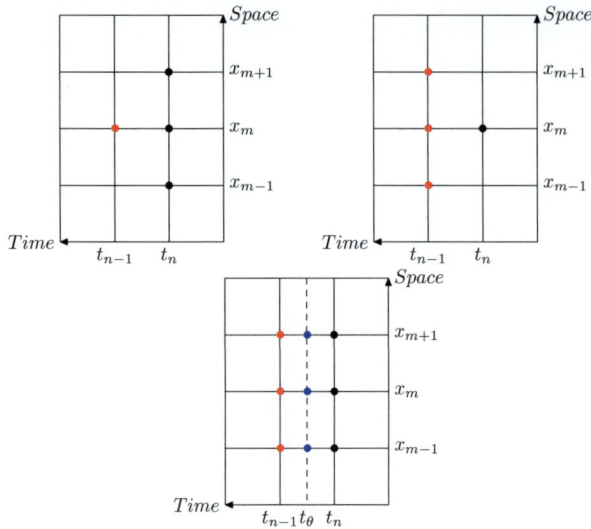

Figure 5.1.: Grid points (black) involved in the Explicit, Implicit and
Weighted Scheme to calculate grid points in red.

The Upwind Scheme

Especially for convection-diffusion PDE's with small or zero diffusion coefficients b and dominating convection coefficients a, standard discretizations like the second order central discretization $\mathcal{D}_x^3$ for the first order derivative show not to be stable. This becomes expressed through spurious oscillations in particular often seen in the region of discontinuous points in the final-time profile $\Phi(.)$ - see also Section 8.3 for examples.

A scheme which restores stability but on the other hand is only of first order space accuracy is called the *Upwind Scheme*. We describe this scheme for a spatial one-dimensional PDE (5.2) with $b \equiv 0$, $r \equiv 0$ and a constant convection coefficient a. This specific PDE is then of hyperbolic type:

$$\frac{\partial u}{\partial t} + a\frac{\partial u}{\partial x} = 0, \quad u(T, x) = \Phi(x) \tag{5.7}$$

This equation has an analytical solution of the form:

$$u(x,t) = \Phi\big(x - a(t - T)\big)$$

On its so-called *Characteristics* $x - a(t - T) = c$ with $c \in \mathbb{R}$ the solution is constant, so we have here at time points $t < T$:

$$u(x,t) \equiv u(x,T) \quad \forall (t,x) \text{ with } \quad x - a(t - T) = c$$

For solving these equations with general coefficients $a(t,x)$ using the Method of Characteristics we refer to [19, pp.84]. The convection coefficient a has the interpretation of being the speed with which the final-time profile $\Phi(x)$ is transported over time along each characteristic. The upwind scheme is simply an explicit scheme consisting of two-point space discretization $\mathcal{D}^2$, which further depends on the sign of the convection coefficient a.

The scheme has an descriptive explanation which is shown by Figure 5.2. In contrast to a central approximation the upwind scheme models the flow of the solution correctly depending on the direction of the characteristics - equivalently to sign of the convection coefficient - which are drafted by the dashed arrows.

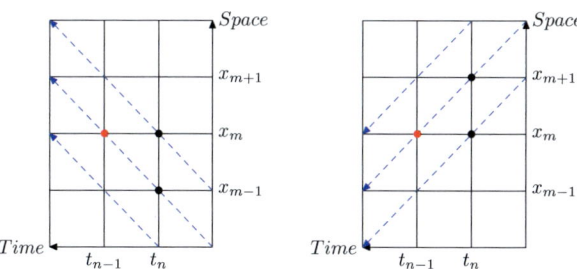

Figure 5.2.: Correct Resolution of the Flow along the Characteristics with $a > 0$ and $a < 0$ by the Upwind Scheme

The scheme equation writes for (5.7):

$$0 = \frac{U_m^n - U_m^{n-1}}{k} + a \begin{cases} \mathcal{D}_x^{2-}, & a > 0 \\ \mathcal{D}_x^{2+}, & a < 0 \end{cases}$$

Rewriting with $\mu := \frac{k}{h}$:

$$\boxed{U_m^{n-1} = U_m^n + \frac{|a| + a}{2}\mu\left(U_m^n - U_{m-1}^n\right) + \frac{|a| - a}{2}\mu\left(U_{m+1}^n - U_m^n\right)}$$

$$(5.8)$$

This scheme equation adds a so-called *Artificial Diffusion* on the original PDE (5.7) as we can rearrange terms and obtain:

$$0 = \frac{U^n - U^{n-1}}{k} + |a|\frac{U_{m+1}^n - U_{m-1}^n}{2h} - \frac{a}{2}h\frac{U_{m-1}^n - 2U_m + U_{m+1}^n}{h^2}$$

So although the upwind scheme is only first order convergent for the original equation (5.7), it can be viewed as a second order scheme for a modified convection-diffusion equation with a non-zero diffusion coefficient:

$$\frac{\partial u}{\partial t} + |a|\frac{\partial u}{\partial x} = \underbrace{\frac{a}{2}h\frac{\partial^2 u}{\partial x^2}}_{\text{Error } O(h)}$$

The introduced diffusion *smears* out the oscillations but has the disadvantage of incorrect approximation behavior, as it - in contrast to the initial PDE - introduces additional damping.

The Lax-Wendroff Scheme

This scheme has a similar behavior as the upwind scheme, as it is explicit, adds artificial diffusion and resolves the flow correctly along the characteristics. The advantage is that this scheme has second order local accuracy in space and time, as it is a direct Taylor Expansion applied to the PDE. The derivation is straightforward and can be found in [5, pp.52]. We give the scheme equation in view of our simplified PDE from (5.7).

$$\frac{U_m^n - U_m^{n-1}}{k} + a\frac{U_{m+1}^n - U_{m-1}^n}{2h} - \frac{a^2}{2}\frac{U_{m-1}^n - 2U_m + U_{m+1}^n}{h^2} = 0$$

Rewriting yields:

$$U_m^{n-1} = U_m^n + \frac{ak}{2h}[U_{m+1}^n - U_{m-1}^n] - \frac{a^2k^2}{2h^2}[U_{m-1}^n - U_m + U_{m+1}^n]$$

$$(5.9)$$

So we end up with the scheme equation:

$$U_m^{n-1} = [1 - \alpha^2]U_m^n - \frac{1}{2}[\alpha(1-\alpha)]U_{m-1}^n + \frac{1}{2}[\alpha(1-\alpha)]U_{m-1}^n$$

where $\alpha := \frac{ak}{h}$. We see that the size of the added artificial diffusion depends also on the size of the convection coefficient but - in contrast to the upwind scheme - in addition on the time step-size k. Furthermore this scheme does produce oscillations - usually still smaller then the centered discretization - as there are cases for which the scheme equation (5.9) does not satisfy the Maximum Principle - see [19, pp.97].

The Effect of Upwind Scheme for the Zero-Coupon Bond Price

In view of the valuation PDE (4.13), we assume spurious oscillations when applying a central discretization $\mathcal{D}_x^3$ in the second space dimension due to the missing diffusion part.

We test here a centered discretization for the second spatial dimension - i.e. the second space variable Y_t - versus the upwind discretization with a large diffusion coefficient η. The following table shows the fixed parameters:

Scheme	Domain	$[-3.0, 3.0] \times [0, 0.31]$
	Grid	$100 \times 150 \times 100$
	Discretization	$\mathcal{D}^3$ (first) + Upwind vs. $\mathcal{D}_x^3$ (second)
PDE	η	0.25
	κ	0.1
Final-Time	$T = 20.0$	$\Phi(T; x, y) \equiv 1.0$

Figure 5.3 shows the bond value $P(t, T; X_t, Y_t)$ at time $t = 0$ in the second space dimension for $X_0 = 0$. The smearing effect of the Upwind Scheme (5.8) is apparent as it resolves the flow correctly, does not produce oscillations as the centered discretization but overshoots the analytical solution.

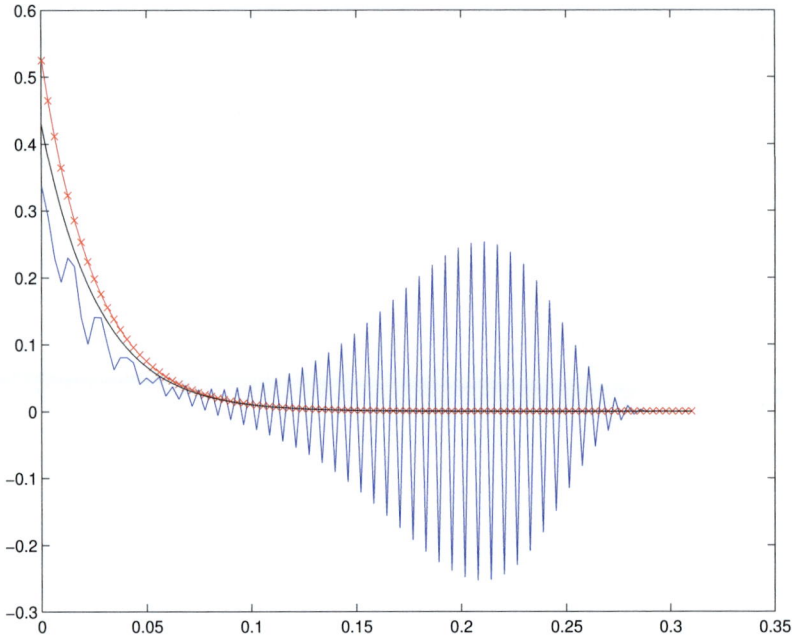

Figure 5.3.: Second Space Dimension of the Zero-Coupon Bond
            $P(t,T;x,y)$ (4.10) with Upwind and Centered Discretiza-
            tion versus Analytical Solution.

Finite Difference Schemes and Markov Chains

In view of the Feynman-Kac PDE (1.34) with which we are solving conditional expectations $V(t, x) = \mathbb{E}[\Phi(X_T) | X_t = x]$ backwards in time, we would like to stress that the above introduced finite difference schemes do also have an probabilistic interpretation.

We discuss this feature here exemplarily for the explicit scheme, where for probabilistic interpretations of implicit and upwind schemes we refer the reader to [21, pp.89] - although in that context finite differences are explained in view of Markov Chains.

The discretized approximations U_m^n correspond to grid approximations of the conditional Expectations which solve (1.34). If we review the explicit scheme (5.5), we see that the approximation of the conditional expectation at time t_{n-1} is simply a weighted sum of the conditional expectations at time t_n. The weights coming from the applied discretization and now redefined as p_i can now be interpreted as probabilities:

$$U_m^{n-1} = p_1 U_{m-1}^n + p_2 U_m^n + p_3 U_{m+1}^n$$

Our numerical scheme equation applied to conditional expectations corresponds directly to the backward equation for Markov Chains (1.5) on a state space with three possible states - correspondingly the probabilities for all other states are zero.

A similar relationship can be recovered for the forward equation (1.4) - see [5, pp.114] for a treatment.

Remark 5.1. *When the time and state space of the underlying Markovian Ito Process is discretized, the explicit finite difference scheme (5.5) applied to the Feynman-Kac PDE resolves into the backward equation for Markov Chains (1.5).*

5.3. Consistency, Stability and Convergence

The *Lax-Richtmeyer Equivalence Theorem* - see [5, p.26] - relates the three notions describing the quality of a Difference Scheme in the sense that if a scheme is consistent - i.e. local accurate - then stability is equivalent to convergence.

Theorem 5.1. *A consistent finite difference scheme for a partial differential equation, for which the initial value problem is well-posed is convergent if and only if it is stable.*

In the following we use the sufficient condition of this theorem, i.e. for convergence analysis we analyze consistency and stability. Firstly we give an overview of the three notions occurring in the equivalence theorem.

Convergence

Lax-Richtmeyer Equivalence Theorem *Convergence* considers the global behavior of the difference scheme in terms of the original *Boundary Value Problem* (BVP) - defined through PDE and Boundary Conditions. The most important question is, how good an approximation U^n, which satisfies our difference scheme, behaves compared to the initially given boundary value problem.

To compare an exact solution $u(t, x)$ to its scheme approximation U_m^n, which is only defined on a discrete time and space grid, we firstly introduce a grid operator G. This operator maps $u(t, x)$ onto the particular chosen grid.

Definition 5.2. *Let $u = u(t, x)$ be an analytical solution of (5.2). The Grid-Operator $G_{k,h}$ maps $u(t, x)$ onto the particular chosen grid $[T_0, T_1]_k \times \Omega_h$.*

$$G_{k,h}u(t, x) := u(t_n, x_m) \quad \forall (t_n, x_m) \in [T_0, T_1]_k \times \Omega_h$$

Using this operator we define the notion of a *Convergent* Scheme for (5.4) following [20, p.45].

Definition 5.3. *A difference scheme of the type (5.4) approximating the partial differential equation (5.2) is a* convergent *scheme of order* (r, s), *if[4] for any* $t \in [T_1, T_2]$, *as* $nk \to t$:

$$\|U_h^{n-1} - G_{k,h}u(t, .)\| = O(k^r) + O(h^s) \quad for \quad k, h \to 0$$

In addition we have to postulate that the solution does depend continuously on the final-time profile given through $\Phi(.)$ This is basically captured by postulating the BVP to be *Well-Posed*. For a detailed analysis of this notion we refer the reader to [5, pp. 169].

Consistency and Local Accuracy

Consistency analysis considers the local behavior of the approximation. The accuracy of the scheme is mainly influenced by the chosen discretization and is described by the *Truncation Error*. This quantity measures the deviation of the discretized solution U_m^n in terms of the exact solution $u(t, x)$. For that to analyze Taylor expansions of the exact solution $u(t, x)$ are replaced into the finite difference scheme which then reveals the local truncation error.

Definition 5.4. *The local* Truncation Error *at* (t_n, x_m) *for a finite difference scheme (5.4) is defined as the residuum, which remains when applying the scheme to the exact solution directly:*

$$\tau_{k,h}(t_n, x_m) := FD_{k,h}G_{k,h}u(t, x)$$

A scheme is said to be consistent *if for* $k, h \to 0$:

$$\tau_{k,h}(t_n, x_m) \to 0$$

A consistent scheme is of local asymptotic order (r, s), *if:*

$$\tau_{k,h}(t_n, x_m) = O(k^r) + \sum_{i=1}^{q} O(h_i^s)$$

---

[4] $\|U^n\|^2 = \left[\sum_{m=0}^{M} h[U_m^n]^2 \right]$

We would like to apply here consistency analysis for (5.2) in one spatial dimension for the general weighted scheme (5.4) with $\theta = \frac{1}{2}$ and central difference operators $\mathcal{D}_x^3$ and $\mathcal{D}_{xx}^3$. The truncation error for the weighted scheme is considered at the intermediate time-points $t_{n-\frac{1}{2}k}$.

Proposition 5.1. *We consider (5.2) with one spatial dimension. For this PDE the weighted scheme (5.4) with $\theta = \frac{1}{2}$ and centered discretizations $\mathcal{D}_x^3$ and $\mathcal{D}_{xx}^3$ is accurate of order $O(k^2) + O(h^2)$.*

Proof. Firstly we write out the Taylor expansions in all directions of the exact solution $u_m^n = u(t_n, x_m)$ of (5.2) at the intermediate time-points $t_{\frac{1}{2}} := t_n - \frac{1}{2}k$. For the time direction we have[5]:

$$u_m^{n-1} = u_m^{n-\frac{1}{2}} - \frac{1}{2}ku_t + \frac{1}{8}k^2 u_{tt} - \frac{1}{16}k^3 u_{ttt} + O(k^4)$$

$$u_m^n = u_m^{n-\frac{1}{2}} + \frac{1}{2}ku_t + \frac{1}{8}k^2 tu_{tt} + \frac{1}{16}k^3 u_{ttt} + O(k^4)$$

For the discretized spatial diffusion operator we have at the chosen midpoint $t_{\frac{1}{2}}$:

$$\mathcal{D}_{xx}^3 u_m^{n-1} = u_{xx}^{n-\frac{1}{2}} + O(h^2) - \frac{1}{2}k\Big[u_{xxt} + O(kh^2)\Big] + \underbrace{\frac{1}{8}k^2\Big[u_{xxtt} + O(kh^2)\Big]}_{=O(k^2)+O(k^3 h^2)}$$

Similarly we have the Taylor expansion in the other direction:

$$\mathcal{D}_{xx}^3 u_m^n = u_{xx}^{n-\frac{1}{2}} + O(h^2) + \frac{1}{2}k\Big[u_{xxt} + O(kh^2)\Big] + \frac{1}{8}k^2\Big[u_{xxtt} + O(kh^2)\Big]$$

In the same manner and by not writing out higher order terms we consider the spatial convection term:

$$\mathcal{D}_x^3 u_m^{n-1} = u_x^{n-\frac{1}{2}} + O(h^2) - \frac{1}{2}k\Big[u_{xt} + O(kh^2)\Big]$$

$$\mathcal{D}_x^3 u_m^n = u_x^{n-\frac{1}{2}} + O(h^2) + \frac{1}{2}k\Big[u_{xt} + O(kh^2)\Big]$$

---

[5]For the ease of notation we use subscripts here for partial derivatives

as well as the spatial reaction term:

$$u_m^{n-1} = u_m^{n-\frac{1}{2}} - \frac{1}{2}hu_x + O(h^2) - \frac{1}{2}k\left[u_t - \frac{1}{2}hu_{xt}h + O(kh^2)\right]$$
$$u_m^{n} = u_m^{n-\frac{1}{2}} + \frac{1}{2}hu_x + O(h^2) + \frac{1}{2}k\left[u_t + \frac{1}{2}hu_{xt} + O(kh^2)\right]$$

Now we can consider the truncation error for the weighted scheme at a glance in combination with our above Taylor expansions and $\mathcal{L}_h = \mathcal{D}_x^3 + \mathcal{D}_{xx}^3 - rx$:

$$\tau_{k,h}(t_{\frac{1}{2}}, x_m) = \frac{u_m^n - u_m^{n-1}}{k} + \theta\mathcal{L}_h u_m^{n-1} + (1 - \theta)\mathcal{L}_h u_m^n$$

$$= \underbrace{u_t + au_x + bu_{xx} - ru}_{=0}\Bigg|_{t=t_{n-\frac{1}{2}k}}$$

$$+ k\left[\frac{1}{2}(1 - \theta) - \frac{1}{2}\theta\right]\left[u_{xt} + u_{xxt} - rhu_x + O(kh)\right]$$

$$+ O(k^2) + O(h^2)$$

$$= k\left[\frac{1}{2} - \theta\right]\left[u_{xt} + u_{xxt} - rhu_x + O(kh)\right] + O(k^2) + O(h^2)$$

If we now set $\theta = \frac{1}{2}$, we will see that the accuracy of the weighted scheme is of second order in time and space. This scheme is usually also known as the *Crank-Nicolson* Scheme. □

It is possible to increase the accuracy when considering higher order terms in the above Taylor expansions. For instance it is shown for the Diffusion Equation (1.8) in [19, pp.28] how to choose θ to obtain local accuracy of order $O(k^2) + O(h^4)$, although this restricts the choice of the step sizes k and h.

Considering now the full implicit scheme we see that its time accuracy is only of first order.

Proposition 5.2. *For the PDE (5.2) the implicit scheme (5.6) with centered discretizations $\mathcal{D}_x^3$ and $\mathcal{D}_{xx}^3$ is of order $O(k) + O(h^2)$.*

Proof. In analogue to the previous proof Taylor expansion of $u(x,t)$ at the point (x_m, t_{n-1}) yields:

$$\tau_{k,h}(t_{n-1}, x_m) = \underbrace{u_t + au_x + bu_{xx} - ru}_{=0}\Big|_{t=t_{n-1}} + O(k) + O(h^2)$$

$\square$

Stability through Fourier Analysis

There are several methods to analyze the stability of a finite difference scheme. Informally a scheme is stable if it does not magnify the resulting local discretization errors at each time step.

We concentrate in the following on the *von-Neumann* Stability Analysis which compares Fourier Modes of the approximation at different time steps postulating that those remain bounded over time. We follow here the treatment in [5, pp.40].

The *von-Neumann* Stability Analysis uses the *Fourier Transformation* $\mathcal{U}^n(z)$ of the scheme approximation U^n which is given by[6]:

$$\mathcal{U}^n(z) = \frac{1}{\sqrt{2\pi}} \sum_{m=-\infty}^{\infty} he^{-imhz}U_m^n \quad \text{where} \ \text{ and } \ z \in [-\frac{\pi}{h}, \frac{\pi}{h}]$$

Thereby it makes use of the fact that solving one step $U^n \to U^{n-1}$ in the finite difference scheme (5.4) is equivalent of multiplying its Fourier Transform of U^n with a so-called *Amplification Factor* λ.

We demonstrate the procedure for a generalized explicit scheme.

---

[6]Here $i := \sqrt{-1}$

Review the particular scheme equation (5.5) which involves grid points U_{m-1}^n, U_m^n and U_{m+1}^n. We then have more general with abstract weighting factors $w_1, w_2, w_3 \in \mathbb{R}$:

$$U_m^{n-1} = w_1 U_{m-1}^n + w_2 U_m^n + w_3 U_{m+1}^n$$

Now the *Fourier Inversion* of $\mathcal{U}^n(z)$ is defined by:

$$U_m^n := \frac{1}{\sqrt{2\pi}} \int_{-\pi/h}^{\pi/h} e^{imhz} \mathcal{U}^n(z)(z) dz$$

Substituting the grid points by their particular inverse Fourier Transforms we obtain for the scheme equation:

$$U_m^{n-1} = \frac{1}{\sqrt{2\pi}} \int_{-\pi/h}^{\pi/h} e^{imhz} \big[w_1 e^{-ihz} + w_2 + w_3 e^{ihz} \big] \mathcal{U}^n(z) dz$$

When comparing this to the Inverse of $\mathcal{U}^{n-1}(z)$ directly:

$$U_m^{n-1} = \frac{1}{\sqrt{2\pi}} \int_{-\pi/h}^{\pi/h} e^{imhz} \mathcal{U}^{n-1}(z) dz$$

Therefore we can deduce - since the Fourier Transform is unique - that proceeding one time step $n \to n-1$ is equivalent to multiplying the Fourier Transform with an amplification factor:

$$\mathcal{U}^{n-1}(z) = \big[w_1 e^{-ihz} + w_2 + w_3 e^{ihz} \big] \mathcal{U}^n(z) =: \lambda(z, k, h) \mathcal{U}^n(z)$$

Now having derived exemplarily the amplification factor for a one-step scheme we give the stability theorem, which roughly postulates that the amplification factors are not allowed to become larger than one. For a proof of the following Theorem see [5, pp.44].

Theorem 5.2. *The scheme (5.4) with constant coefficients is stable if and only if there exists a constant K and $k_0 > 0$ and $h_0 > 0$ such that:*

$$|\lambda(\varphi, k, h)| \leq 1 + Kk$$

$$\forall \varphi \in [-\pi, \pi], \ \forall 0 < k \leq k_0, \ \forall 0 < h \leq h_0.$$

Practically we can simplify the above procedure by making the following replacement in the particular scheme equation and solving for λ:

$$U_m^n = \lambda^n e^{im\varphi} \text{ with } \varphi := hz \quad \text{therefore} \quad \varphi \in [-\pi, \pi]$$

In view of the difference operators $\mathcal{D}$ from Table 5.1 a direct application of the above replacement yields the following expressions in terms of their Fourier Modes.

Operator	Fourier Mode
$\mathcal{D}_x^{2-} U_m$	$\frac{1}{h}\left[e^{-i\varphi} - 1\right]e^{im\varphi}$
$\mathcal{D}_x^{2+} U_m$	$\frac{1}{h}\left[e^{i\varphi} - 1\right]e^{im\varphi}$
$\mathcal{D}_x^3 U_m$	$\frac{1}{h}\left[i\sin\varphi\right]e^{im\varphi}$
$\mathcal{D}_{xx}^3 U_m$	$\frac{2}{h^2}\left[\cos\varphi - 1\right]e^{im\varphi}$

As one additional remark, when comparing these factors to the *eigenvalues* of the corresponding tridiagonal matrices, we see that those are closely related - see [20, pp.52]. So the requirement for the amplification factor to satisfy the above inequality is comparable to the requirement that the eigenvalues of the matrix sequence to be solved are lower equal to one.

We now derive the amplification factor for a specific PDE (5.2) with one spatial dimension, which builds the basis for PDE's in more dimensions. For some more examples we refer to [5, pp.43].

Proposition 5.3. *Consider the PDE with one spatial dimension and with $a, b, r \in \mathbb{R}$ and $b \geq 0$, $r \geq 0$:*

$$\frac{\partial u}{\partial t} + a\frac{\partial u}{\partial x} + b\frac{\partial^2 u}{\partial x^2} = ru$$

For this PDE the amplification factor of the general weighted scheme (5.4) with central differences $\mathcal{D}_x^3$ and $\mathcal{D}_{xx}^3$ is given by:

$$\lambda = \frac{1 - \left[(1-\theta)kr + (1-\theta)2b\mu(1-\cos\varphi)\right] + (1-\theta)ia\mu h\sin\varphi}{1 + \left[\theta kr + \theta 2b\mu(1-\cos\varphi)\right] - \theta ia\mu h\sin\varphi}$$

$$(5.10)$$

where $\mu := \frac{k}{h^2}$ and $\varphi \in [-\pi, \pi]$.

Proof. This is a straightforward calculation from the rewritten scheme (5.4):

$$U_m^{n-1} - \theta k\mathcal{L}_h U_m^{n-1} = U_m^n + (1-\theta)k\mathcal{L}_h U_m^{n-1}$$

Replacing the difference operators $\mathcal{D}$ with their Fourier Modes from the above table and solving for λ leads directly to factor (5.10). □

From that we can deduce directly the following two propositions:[7]

Proposition 5.4. *The implicit scheme (5.6) with $\theta = 1$ is unconditionally stable.*

Proof.

$$|\lambda| = \frac{1}{\sqrt{(1 + kr + 2b\mu(1-\cos\varphi))^2 + a\mu h\sin^2\varphi}} \leq \frac{1}{1+kr}$$

As we assumed $r \geq 0$, so $|\lambda| \leq 1$ therefore the scheme is stable. □

---

[7] Norm of a complex number $z = x + iy$: $|z| = \sqrt{z\bar{z}} = \sqrt{x^2 + y^2}$

Proposition 5.5. *The weighted scheme (5.4) with $\frac{1}{2} \leq \theta < 1$ is uncon-ditionally stable.*

Proof. We introduce $\vartheta := 1 - \theta$, so we have that $\theta \geq \vartheta$:

$$|\lambda| = \frac{\sqrt{\left(1 - \left[\vartheta kr + \vartheta 2b\mu(1 - \cos\varphi)\right]\right)^2 + (\vartheta a\mu h \sin\varphi)^2}}{\sqrt{\left(1 + \left[\theta kr + \theta 2b\mu(1 - \cos\varphi)\right]\right)^2 + (\theta a\mu h \sin\varphi)^2}} \leq \frac{|1 - \vartheta kr|}{|1 + \theta kr|} \leq 1$$

$\square$

We remarked that the explicit scheme (5.5) usually does have severe stability restrictions. We would like to demonstrate this for the Heat Equation (1.8).

Proposition 5.6. *Consider the Heat Equation:*

$$\frac{\partial u}{\partial t} = \frac{\partial^2 u}{\partial x^2}$$

for which the explicit scheme (5.5) is only stable, if

$$\mu \leq \frac{1}{2}$$

Proof. The amplification factor (5.10) writes in this simplified case with $\theta = 0$:

$$|\lambda| = |1 + 2\mu(1 - \cos\varphi)| \leq |1 - 4\mu| \overset{!}{\leq} 1$$

Following [5, p.119] we postulate here the amplification factor to be smaller equal to one. This implies $\mu \leq \frac{1}{2}$. $\square$

That last proposition is useful in view of an additional reason. We explained in the last section how a explicit difference scheme corresponds to the backward equation (1.5) for Markov Chains. In this context, the above restriction has the interpretation, that the transition probabilities p_i lie always in the interval $[0, 1]$.

5.4. Alternating Direction Implicit Schemes (ADI)

The general Idea

We describe the general idea using PDE (5.2) in two space dimensions.

Applying a weighted scheme combined with a centered discretization for both space differential operators would require solving a linear equation system of dimension $P \times Q$ with $P := M_1 + 1$ and $Q := M_2 + 1$ for each of the N time steps.

Usually these matrices are sparse but do not have a tridiagonal structure. As we are not able to apply the Thomas Algorithm, we have to make use of iterative solvers which increase the asymptotic workload to at least to $O\big((P \times Q)^r\big)$ with $r > 1$.

The essential idea developed by Peaceman and Rachford (1955) was to solve a splitted scheme equation for each single spatial operator separately successively over additional intermediate time steps.

Now when applying a discretization involving three grid points this implies to solve a sequence of tridiagonal systems. Therefore this procedure decreases the complexity to $O(P \times Q)$ for each time step.

The other advantage is that the local accuracy order remains the same, since the splitting is performed by adding terms of higher order than the accuracy of the original scheme.

We now begin to analyze the original Peaceman & Rachford Scheme and afterwards the Craig & Sneyd Scheme applicable to PDE's with more space dimensions.

The original Peaceman-Rachford (PR) Scheme

We follow here [19, pp.60]. Consider (5.4) with $\theta = \frac{1}{2}$ and $q = 2^8$:

$$\left[1 - \frac{1}{2}k\bar{D}_x - \frac{1}{2}k\bar{D}_y\right]U^{n-1} = \left[1 + \frac{1}{2}k\bar{D}_x + \frac{1}{2}k\bar{D}_y\right]U^n$$

Adding the higher order terms $\frac{1}{4}k^2\bar{D}_x\bar{D}_y$ we can factor out and obtain the following:

$$\left[1 - \frac{1}{2}k\bar{D}_x\right]\left[1 - \frac{1}{2}k\bar{D}_y\right]U^{n-1} = \left[1 + \frac{1}{2}k\bar{D}_x\right]\left[1 + \frac{1}{2}k\bar{D}_y\right]U^n$$

$$(5.11)$$

Introducing the intermediate value $U(\bar{t})$ at time-step $\bar{t} := t^{n-\frac{1}{2}k}$ this factorized equation can now be split up in two steps:

Algorithm 5.1.

$$\text{Step I} \qquad \left[1 - \frac{1}{2}k\bar{D}_x\right]U(\bar{t}) \quad = \quad \left[1 + \frac{1}{2}k\bar{D}_y\right]U^n$$

$$\text{Step II} \qquad \left[1 - \frac{1}{2}k\bar{D}_y\right]U^{n-1} \quad = \quad \left[1 + \frac{1}{2}k\bar{D}_x\right]U(\bar{t}) \qquad (5.12)$$

Proof. We mainly use that the matrix operator $[1 - \frac{1}{2}k\bar{D}_x]$ commutes with $[1 + \frac{1}{2}k\bar{D}_x]$. Reversely working we multiply the second step with $[1 - \frac{1}{2}k\bar{D}_x]$ and interchange the operators on the right hand side. We then find out that it corresponds exactly to the left hand side of the first step. Therefore we can replace the left hand side of the first step with the modified left hand side of the second step and arrive at the above unsplit form. □

---

[8]Note, that we are using a somewhat more abstract (scalar) notation for the derivation, although our difference operators are matrices.

It is shown in the given reference that the additionally included terms are of the same order as the truncation error which is of second order accuracy in space and time $O(k^2) + O(h_1^2) + O(h_2^2)$ - comparable to the one dimensional weighted scheme (for this result see also [20, p.167]).

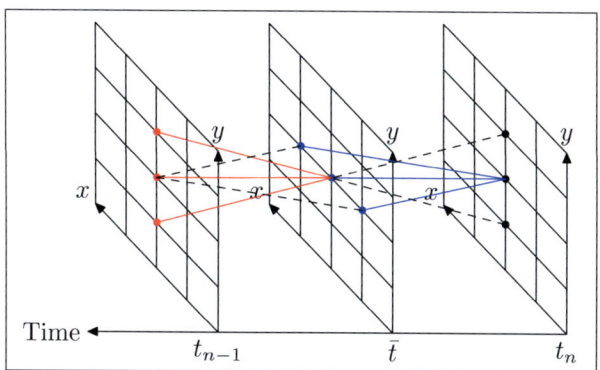

Figure 5.4.: Alternating Implicitness of the PR-Scheme

Figure 5.4 demonstrates the procedure. The dashed lines show which operators are included explicitly - i.e. on the right hand side of the scheme equation: Firstly we solve implicitly for the first spatial dimension while treating the second spatial operator explicitly and afterwards - *alternating the implicitness* - apply the reverse in the second step.
In conclusion: We have a weighted scheme with $\theta = \frac{1}{2}$, where the weighting is applied to different spatial directions in each step.

Proposition 5.7. *Consider (5.2) in two spatial dimensions:*

$$\frac{\partial u}{\partial t} + \sum_{i=1}^{2} \left[a_i \frac{\partial u}{\partial x} + b_i \frac{\partial^2 u}{\partial x^2} \right] = ru$$

For this PDE, the above scheme (5.12) is unconditional stable.

Proof. Starting with the unsplit scheme equation (5.11) we replace the difference operators with their according amplification factors. After grouping the operators corresponding to their spatial direction, we can make use of the previously derived one-dimensional amplification factor (5.10). Applying those here denoted as λ_{x_1} for the first and λ_{x_2} for the second spatial direction. We see that we have:

$$|\lambda| = |\lambda_{x_1}||\lambda_{x_2}|$$

We have already proven in Proposition 5.5 that for the particular factors in one dimension holds: $|\lambda_{x_i}| \leq 1$. Therefore we also have $|\lambda| \leq 1$ □

A general weighted scheme - Craig & Sneyd (CS)

One of the main problems with ADI schemes occurs, when mixed partial derivatives are present. Then it is not that intuitive to solve for each single spatial dimension separately since the spatial differential operator is not a sum of orthogonal operators anymore. This problem was overcome in the following scheme proposed in [18] which has some advantages compared to the traditional one:

⋄ It is applicable to PDE's with multiple spatial dimensions

⋄ The scheme can handle mixed derivatives $\frac{\partial}{\partial x_i \partial x_j}$

⋄ It incorporates other previously developed ADI schemes, e.g. $q = 3$ and $\theta = 1$ yields the Douglas-Rachford Scheme, discussed e.g. in [19, pp.68].

As an additional note on mixed derivative terms although they can be handled by the scheme, this reduces in the simple version to first order time accuracy $O(k)$, since those operators are treated only explicitly. Therefore also an *Iterated* Predictor-Corrector version of the scheme has been developed - see [18, p.343] - which time-centers the mixed derivative operators and restores second order time accuracy with $\theta = \frac{1}{2}$. In view of the valuation PDE (4.13) respectively (5.2) we now write the *simple* scheme equation in unsplit form in the absence of mixed derivative terms. For the more general scheme we refer to [18]. The scheme equation writes again in time-backward and unsplit form:

Algorithm 5.2. *The scheme equation for (5.2) with q spatial dimensions writes:*

$$AU^{n-1} = [A + B]U^n \qquad (5.13)$$

where the operators A and B - respectively matrices on the grid - are given by:

$$A := \prod_{i=1}^{q} \left[1 - \theta k \bar{D}_{x_i}\right] \quad and \quad B := k \sum_{i=1}^{q} \bar{D}_{x_i}$$

The parameter θ defines the implicitness of the method.

Algorithm 5.3. *The CS Scheme (5.13) writes in splitted form for two spatial dimensions (q = 2):*

Step I $\left[1 - \theta k \bar{D}_x\right]U(\bar{t}) \quad = \quad \left[1 + (1 - \theta)k\bar{D}_x + k\bar{D}_y\right]U(t_n)$

Step II $\left[1 - \theta k \bar{D}_y\right]U(t_{n-1}) \quad = \quad U(\bar{t}) - \theta k\bar{D}_y U(t_n) \qquad (5.14)$

Proof. We would like to show in the case of two spatial dimensions how this scheme can be derived from our general weighted scheme (5.4). This writes with difference operators $\bar{D}_x$ and $\bar{D}_y$ in two spatial dimensions:

$$\left[1 - \theta k\bar{D}_x - \theta k\bar{D}_y\right]U^{n-1} = \left[1 + [1 - \theta]k\bar{D}_x + [1 - \theta]k\bar{D}_y\right]U^{n-1}$$

Now we add the term $\theta^2 k^2 \bar{D}_x \bar{D}_y$ on both sides which yields:

$$\left[1 - \theta k\bar{D}_x\right]\left[1 - \theta k\bar{D}_y\right]U^{n-1} =$$
$$= \left[1 + [1 - \theta]k\bar{D}_x + k\bar{D}_y - [1 - k\theta\bar{D}_x]k\theta\bar{D}_y\right]U^n$$

Rewriting yields the modified unsplit scheme equation:

$$\left[1 - \theta k\bar{D}_x\right]\left[\; [1 - \theta k\bar{D}_y]U^{n-1} + [k\theta\bar{D}_y]U^n\right] =$$
$$= \left[1 + [1 - \theta]k\bar{D}_x + k\bar{D}_y\right]U^n \qquad (5.15)$$

Introducing an intermediate value $\bar{U}$ - at an intermediate time-point $\bar{t}$ - which we define as:

$$\bar{U} := U(\bar{t}) = [1 - \theta k\bar{D}_y]U^{n-1} + [k\theta\bar{D}_y]U^n$$

yields the second step of (5.14)

$$[1 - \theta k \bar{D}_y] U^{n-1} = \bar{U} - k\theta \bar{D}_y U^n$$

where the first step is recovered by replacing $\bar{U}$ on the left hand side of the above unsplit scheme equation (5.15).   □

Note that we started from the general scheme equation (5.4) which has dimension $P \times Q$. Due to the introduction of the intermediate step with values $\bar{U}$ we could break this dimension down into a sequence of matrices of dimension $P := M_1 + 1$ and $Q := M_2 + 1$. To be more accurate we now have Q linear systems of dimension P to solve in the first step which is followed by a sequence of P linear systems of dimension Q in the second step.
The splitting for three spatial dimensions would be done similarly in three steps.

The alternating implicitness is not that obvious in the two-dimensional CS-Scheme as shown in Figure 5.5: We include all additional inactive operators explicitly in the first step. In the second step - in contrast to the PC Scheme (5.12) - we do the weighting only in one direction - therefore we can extend this procedure to multiple spatial dimensions.

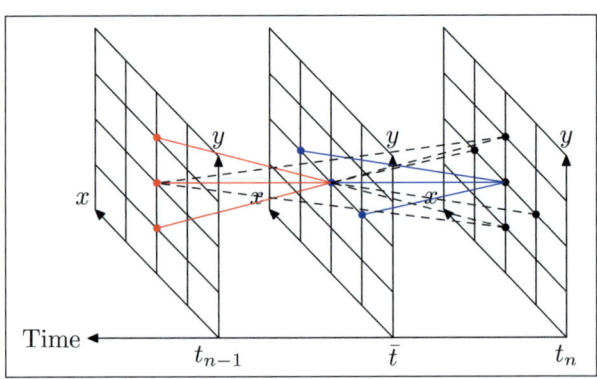

Figure 5.5.: Splitting Procedure in the 2D CS-Scheme with $0 < \theta < 1$

The ADI-Scheme applied to the Valuation PDE (4.13)

We rewrite the valuation PDE (4.13) in the sense of (5.2):

$$\frac{\partial V}{\partial t} + D_x V + D_y V = 0$$

with:

$$D_x := a_1(t, x, y)\frac{\partial}{\partial x} + b_1(t, x, y)\frac{\partial^2}{\partial x^2} - r(t, x)$$

$$D_y := a_2(t, x, y)\frac{\partial}{\partial y}$$

and where we define the coefficients from (4.13) as:

$$a_1 := a_1(t, x, y) = y - \kappa x, \quad b_1 := b_1(x, y, z) = \frac{1}{2}\eta^2(t, x, y),$$

$$r := r(t, x) = f(0, t) + x, \quad a_2 := a_2(t, x, y) = \eta^2(t, x, y) - 2\kappa y$$

How the scheme (5.14) is applied in detail for this particular PDE with time and space dependent coefficients and what kind of additional considerations apply, will now be discussed in detail.

Discretization Operators

Several discretization methods can be applied to the first derivative and the second derivative. Note, since we work on the grid now, the time and space dependence of our coefficients resolves into a dependence on the particular grid point, which itself is specified by the number of steps n, m_1, m_2. As an example when using the discretizations $\mathcal{D}_x^3$ and $\mathcal{D}_{xx}^3$ at the grid-point (t_n, x_{m_1}, y_{m_2}) this constitutes one single row in our operator matrix $\bar{D}_x$. When solving only in the first dimension this vector writes- i.e. m_2 remains fixed - at the m_1-th grid point after arranging coefficients:

$$\bar{D}_x U^n \Big|_{m_1} = \underbrace{(-a_1\nu_x + a_{11}\nu_{xx})}_{:=A_{m_1}} U^n_{m_1-1}$$

$$+ \underbrace{(-r - 2b_1\nu_{xx})}_{:=B_{m_1}} U^n_{m_1} + \underbrace{(a_1\nu_x + a_{11}\nu_{xx})}_{:=C_{m_1}} U^n_{m_1+1} \qquad (5.16)$$

with the factors $\nu_x := \frac{1}{2h_1}$ and $\nu_{xx} := \frac{1}{h_1^2}$. In analogue we can discretize
the other operator D_y at every single grid point m_2 when solving for
the second spatial dimension.

The First Scheme Step in Detail

We now analyze the first step in scheme equation (5.14) in a more
detailed matrix form. Considering this first step which solves implicitly
only for the first operator we have to solve a linear system at all grid
points in the fixed second dimension M_2. This means that for $0 \leq m_2 \leq$
M_2 fixed, the LHS of (5.14) can be written as a set of linear equations.
In combination with the difference operator $\bar{D}_x$ when applied to all
grid points $0 \leq m_1 \leq M_1$ in the sense of (5.16) becomes a matrix of
tridiagonal shape.
The right hand side is directly calculated from values U^n of the previous
time step t_n. Let $I \in \mathbb{R}^P \times \mathbb{R}^P$ denotes the identity matrix. Then the
scheme equation writes:

$$\left[I - \theta \bar{D}_x\right]\bar{U}_{m_2} = \left[I + k(1-\theta)\bar{D}_x + \underbrace{k\bar{D}_y}_{(1)}\right]U^n_{m_2} \tag{5.17}$$

With

$$\bar{D}_x = \begin{bmatrix} B_0 & C_0 & & & \\ A_1 & B_1 & C_1 & & \\ & & \ddots & & \\ & & A_{M_1-1} & B_{M_1-1} & C_{M_1-1} \\ & & & B_{M_1} & C_{M_1} \end{bmatrix} \qquad U^n_{m_2} = \begin{bmatrix} U^n_{0,m_2} \\ \\ \vdots \\ \\ U^n_{M_1,m_2} \end{bmatrix}$$

Firstly consider the matrix operator $\bar{D}_x$ whose shape is determined by
the chosen spatial discretization. For the first and last equation concerning the boundary grid points U_0 and U_{M_1} the problem occurs that
a discretization involving three grid points would also include so-called
ghost points - for instance U_{-1} - which lie outside the domain. This can
be circumvented by an ably integration of either analytic or numerical
boundary conditions. These are either directly provided by the original
problem formulation or - as in our case - have to be assumed or derived

from the model environment. This yields a modified first and last row in the above matrix consisting only of at most two non-zero values. For further details see Section 5.5.

Secondly looking at the right hand side of scheme equation (5.17) for each fixed grid-point in the second space dimension term (1) representing the inactive operator simply resolves into a sum of vectors which become added on the right hand side. For instance when using a centered discretization in the second space dimensions, the vectors $U^n_{m_2-1}$, $U^n_{m_2}$, $U^n_{m_2+1}$ would become added after being multiplied by appropriate discretization factors.

The procedure applied for the second step in (5.14) is essentially the same. Note that on the right hand side we use values from the intermediate $\bar{t}$ *and* previous time step t_n.

Consistence and Stability in the case $q = 2$

In [18] only diffusion terms with constant coefficients - additionally mixed derivatives - were considered. We like to show that local second order accuracy also holds for (5.2) with $q = 2$ and the scheme is unconditional stable.

The scheme for $q = 2$ writes in unsplit form and with the use of difference operators:

$$\left[1 - \theta k \bar{D}_x\right]\left[1 - \theta k \bar{D}_y\right] U^{n-1}$$
$$= \left[\left[1 - \theta k \bar{D}_x\right]\left[1 - \theta k \bar{D}_y\right] + k \bar{D}_x + k \bar{D}_y\right] U^n \qquad (5.18)$$

Proposition 5.8. *For (5.2) with $q = 2$ the scheme (5.14) with $\theta = \frac{1}{2}$ is accurate of order $O(k^2) + O(h_1^2) + O(h_2^2)$*

Proof. There is not much difference compared with the proof for one spatial dimension. Rewriting yields:

$$\left[1 - \theta k \bar{D}_x - \theta k \bar{D}_y + \theta^2 k^2 \bar{D}_x \bar{D}_y\right] U^{n+1}$$
$$= \left[1 + (1 - \theta)k \bar{D}_x + (1 - \theta)k \bar{D}_y + \theta^2 k^2 \bar{D}_x \bar{D}_y\right] U^n$$

A straightforward consideration concerning the mixed terms $\bar{D}_x \bar{D}_y$ shows that these are of the same accuracy order to be proven[9]. Therefore we have for truncation error in combination with our findings for the one-dimensional case - see Proof of Proposition 5.1:

$$\tau_{k,h}(t_{n-\frac{1}{2}k}, x_m) = \frac{u_m^{n+1} - u_m^n}{k} + \theta \mathcal{L}_h u_m^{n-1} + (1-\theta)\mathcal{L}_h u_m^n + O(h_1^2) + O(h_2^2)$$

$$= \underbrace{u_t + D_x u + D_y u}_{=0}\bigg|_{t=t_{n-\frac{1}{2}k}}$$

$$+ k[\theta - \frac{1}{2}]\Big[u_{xt} + u_{xxt} - rhu_x + O(kh_1)\Big]$$

$$+ k[\theta - \frac{1}{2}]\Big[u_{yt} + u_{yyt} + O(kh_2)\Big] + O(k^2) + O(h_1^2) + O(h_2^2)$$

Setting $\theta = \frac{1}{2}$ we obtain the desired accuracy. $\square$

Proposition 5.9. *For the PDE (5.2) with $q = 2$ the scheme (5.14) with $\theta = \frac{1}{2}$ is unconditional stable.*

Proof. For the case $\theta = \frac{1}{2}$ the scheme (5.14) reminds one of the PC-Scheme, in particular only the splitting is done in a slightly different way. The unsplit scheme equation (5.18) now becomes:

$$\Big[1 - \frac{1}{2}k\bar{D}_x\Big]\Big[1 - \frac{1}{2}k\bar{D}_y\Big]U^{n-1} = \Big[1 + \frac{1}{2}k\bar{D}_x\Big]\Big[1 + \frac{1}{2}k\bar{D}_y\Big]U^n$$

which is exactly the unsplit equation of the PC-Scheme for which we already proved unconditional stability (Proposition 5.7). $\square$

Fur a further detailed stability analysis of the scheme in higher spatial dimensions we refer to [18].

---

[9]It is probably not that straightforward to see, but can be found on [20, p.167]

Computational Complexity and Performance Testing

Proposition 5.10. *The scheme (5.14) involves a computational asymptotic complexity of order $O(N \times P \times Q)$ with N the number of time steps, $P := M_1 + 1$ and $Q := M_2 + 1$.*

Proof. We remarked in Section 5.1 that applying the Thomas Algorithm involves $8P$ Floating Point Operations per tridiagonal system of dimension P. For each of the N time steps in (5.14) we now solve Q tridiagonal systems of dimension P in the first step - i.e. $8PQ$ operations - followed by P systems of dimension Q. So we have $16PQ$ operations (only for solving the linear systems) which leads to the desired asymptotic complexity. □

The following CPU times [sec] for several space discretizations were obtained for the valuation PDE (4.13) solved by the scheme (5.14) (Intel Celeron Processor with 2.80 GHz, MS Visual Studio.Net 2003 Compiler).

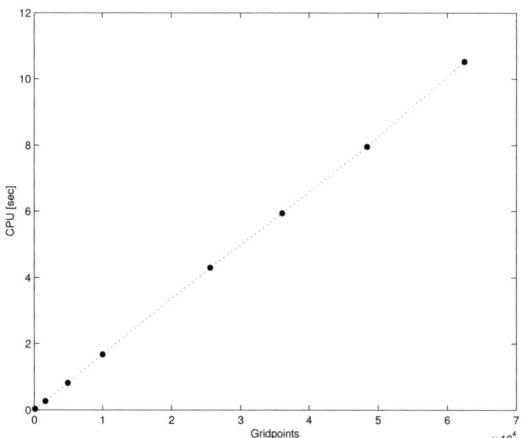

Grid	CPU
10×10	0.1
40×40	0.3
70×70	0.8
100×100	1.7
160×160	4.3
190×190	6.0
220×220	8.0
250×250	10.5
400×400	29.0

One can nicely observe the linear computational complexity which we proved in Proposition 5.10. At each time steps there are $16PQ$ operations so we have a factor ≈ 1.6 in correspondence with the obtained CPU times.

Extensions to cope with the valuation PDE

We list some features which are specific for the valuation PDE (4.13). For more observations and implementation results we refer to later Chapters.

⋄ In Figure 5.3 we observed oscillations when using central discretization D_x^3 in the second spatial dimension, due to the fact that the diffusion term is absent and therefore a wrong resolution of the flow.We therefore add artificial diffusion by applying either the *Discretization* of the Upwind Scheme (5.8) or Lax-Wendroff Scheme (5.9) to the second space dimension in the used ADI Scheme (5.14).

⋄ To obtain maximal local accuracy in the time direction we choose the weighting parameter $\theta = \frac{1}{2}$.

⋄ As the coefficients are also time-dependent it becomes important, where we define the mid-point to calculate the intermediate values. The most intuitive one is simply $\bar{t} = t_n - \frac{1}{2}k$.

⋄ Importantly the market yield curve data at the valuation time point $f^M(0,t)$ determine the reaction term $r(t,x)$ in (4.13). We used 100 points per year on the market discount curve (2.2) from which we calculate time-discrete approximations of the continuous forward rates (2.10) through their corresponding term rates (2.9). If we want to solve the PDE on an arbitrary time grid, we apply a *linear interpolation* between the corresponding discount factors to calculate the approximation of the forward rate between two time steps:

Let $P(0, T_1)$ and $P(0, T_4)$ be two explicitly known discount factors. Assume that we we want to have the forward rate for $P(0, T_2)$ and $P(0, T_3)$ with $T_1 \leq T_2 \leq T_3 \leq T_4$ in the following sense:

$$P(0, T_i) = P(0, T_1) + \frac{T_i - T_1}{T_4 - T_1}[P(0, T_4) - P(0, T_1)] \quad i = 2, 3$$

Afterwards we have the approximation of the forward rate with

(2.9) at time step $T_3 \to T_2$:

$$F(0, T_2, T_3) = \frac{1}{T_3 - T_2} \frac{P(t, T_2) - P(t, T_3)}{P(t, T_3)}$$

5.5. Treatment of Boundary Conditions

We mentioned earlier that boundary conditions are either provided by the problem formulation or have to be assumed for computational reasons. From our view the only boundary condition which is explicitly provided is the final-time condition $V_T = \Phi(x)$ and usually corresponds to the payoff of the financial product to be valued. We are left with the problem that the Feynman-Kac Theorem is valid on entire $\mathbb{R}^q$. As we have to specify a bounded space domain, we are faced with the problems *where* to truncate the domain and *how* to approximate the boundary behavior of the PDE at those domain boundaries. We investigated two main types of boundary conditions:

$\diamond$ *Analytical Conditions*

$$\text{Dirichlet Condition:} \qquad u = g(t,x,y) \qquad (5.19)$$

$$\text{Neumann Condition:} \qquad \frac{\partial u}{\partial x} = g(t,x,y) \qquad (5.20)$$

$$\text{Robbins Condition:} \qquad p\frac{\partial u}{\partial x} + qu = g(t,x,y) \qquad p,q \in \mathbb{R}$$
$$(5.21)$$

where we extend the treatment of [22, pp.48] to the convection-diffusion PDE (5.2).

$\diamond$ A *Numerical boundary condition*[10] which is basically a *linear extrapolation* at the particular boundary. This condition postulates:

$$\frac{\partial^2 u}{\partial x^2} = 0 \qquad (5.22)$$

Although all boundary conditions have been implemented we spend here only a small subsection on the analytical ones, as they are difficult to apply in view of our purposes. It has turned out that the linear extrapolation approach provides a flexible *generic* framework in view of the Feynman-Kac PDE and the purpose of product valuation.

---

[10]For further numerical boundary conditions we refer to [5, pp. 74]

Analytical Boundary Conditions

We have to use a discretization method for the occurring first derivatives, which in combination with our scheme results into extended formulas for our linear systems. We provide these formulas explicitly, since they can be directly used for the implementation - see also Chapter 9. As we did in the previous section, we explain the procedure for the first scheme step and for the lower domain value U_0, which has to satisfy scheme equation and the analytical boundary condition.

Dirichlet Condition

This is the easiest condition to implement. Solving for instance the first step, we would have at the intermediate time point $U_0^n = g(t_n, x_0, y_{m_2})$ - in turn we have for the intermediate value $\bar{U}_0$ the first row of (5.17) with $B_0 = 1$ and $C_0 = 0$:

$$\bar{U}_0 = g(\bar{t}, x_0, y_{m_2})$$

This approach may lower the order of accuracy at the boundaries, because we do not include the scheme equation (5.17) which has to be also satisfied at $\bar{U}_0$. How this can be resolved more accurately see [20, pp.169].

Neumann and Robbins Condition

Naturally we firstly also discretize the occurring first order derivative in the boundary condition. To have second order accuracy at the boundaries we use $\mathcal{D}_x^3$ from Table 5.1:

$$p\nu_x(U_1^n - U_{-1}^n) + qU_0 = g(t_n, x_0)$$
$$p\nu_x(U_1^{n-1} - U_{-1}^{n-1}) + qU_0^{n-1} = g(t_{n-1}, x_0)$$

with $\nu_x = \frac{1}{2h_1}$. These discretized conditions involve grid points U_{-1}^n and U_{-1}^{n-1}, which are located outside the domain. Therefore they have to be eliminated which is achieved through a *linear combination* of the system equation for U_0, which corresponds to the first row in (5.17) and with the two above boundary equations for both time steps n and $n-1$ [11]:

$$-\theta A_0 U_{-1}^{n-1} + (1 - \theta B_0)U_0^{n-1} - \theta C_0 U_1^{n-1} =$$
$$(1 - \theta)A_0 U_{-1}^n + (1 + (1 - \theta)B_0)U_0^n + (1 - \theta)C_0 U_1^n \qquad (5.23)$$

Here we use the notations A_0, B_0, C_0 from the previous section. Straightforward calculation reveals the *modified* first row of the linear system:

$$\left[1 - \theta B_0 - q\frac{\theta A_0}{p\nu_x}\right]\mathbf{U_0^{n-1}} + \left[-\theta A_0 - \theta C_0\right]\mathbf{U_1^{n-1}} =$$
$$\left[1 + (1 - \theta)B_0 + q\frac{(1 - \theta)B_0}{p\nu_x}\right]\mathbf{U_0^n} + \left[(1 - \theta)C_0 + (1 - \theta)A_0\right]\mathbf{U_1^n}$$
$$- \frac{\theta A_0}{p\nu_x}g(t_{n-1}, x_0) + \frac{(1 - \theta)A_0}{p\nu_x}g(t_n, x_0) \qquad (5.24)$$

In analogue we obtain the modified equation for the upper bound corresponding to the last row in our system equation. Here U_{M+1} has to be eliminated which yields:

$$\left[-\theta C_M - \theta A_M\right]\mathbf{U_M^{n-1}} + \left[1 - \theta d_M + q\frac{\theta C_M}{p\nu_x}\right]\mathbf{U_M^{n-1}} =$$
$$\left[(1 - \theta)A_M + (1 - \theta)C_M\right]\mathbf{U_{M-1}^n} + \left[1 + (1 - \theta)B_M - q\frac{(1 - \theta)C_M}{p\nu_x}\right]\mathbf{U_M^n}$$
$$+ \frac{\theta C_M}{p\nu_x}g(t_{n-1}, x_M) + \frac{(1 - \theta)C_M}{p\nu_x}g(t_n, x_M) \qquad (5.25)$$

We have to distinct the case when $A_0 = 0$. Then a linear combination is not possible, therefore we replace the central differences with first order differences $\mathcal{D}_x^{2-}$ and $\mathcal{D}_x^{2+}$ respectively.

---

[11] We do not make a distinction here, which intermediate time step we are considering in the ADI Scheme.

Numerical Boundary Conditions

Postulating the second derivative of the solution to vanish at the boundaries can be incorporated quite simply into the scheme. With the standard discretization for the second order derivative $\mathcal{D}^3_{xx}$ from Table 5.1 we can discretize (5.22). Thereby we obtain that the points located outside the domain can be expressed in terms of grid points within the domain at the lower and upper boundaries: Writing a Taylor Expansion in combination with assumption (5.22) yields:

$$U_{M+1} = U_M + \frac{\partial U}{\partial x} h$$

Using $\mathcal{D}^3_x$ for first the partial derivative we obtain:

$$U_{M+1} = 2U_M - U_{M-1}$$

Similarly we get $U_{-1} = 2U_0 - U_1$. So the ghost points outside the domain are a *linear extrapolation* of values inside the domain. With these expressions we can eliminate U_{-1} and U_{M+1} in our system equation (5.23).

Linear Extrapolation in view of Feynman-Kac

Applying the numerical boundary condition (5.22) shows to be a very generic approach for handling boundary conditions in our case, since the Feynman-Kac Theorem 1.4 does not provide any explicit information about space boundary conditions.

In view of the fact that the valuation PDE describes the dynamics of the conditional expectations $V(t, x)$ backwards in time, we have to find regions where this functional behavior - when assumed to be linear - introduces an error at the boundaries, which is negligible.

So after solving the problem *how* to choose the boundary conditions, the question remains, *where* we are allowed to assume linear behavior, such that this assumption does not distort the values from the domain, we are interested in.

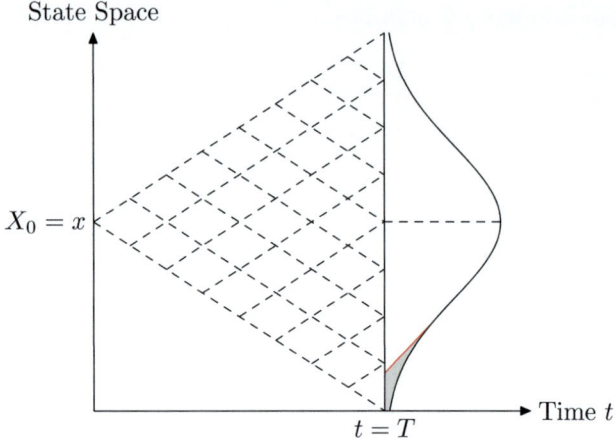

State Space

$X_0 = x$

$t = T$

Time t

Figure 5.6.: Numerical Error of Linear Extrapolation at the boundaries might introduce an error (grey area) in view of the evaluation of $V(0,x) = \mathbb{E}[\Phi(X_T)|X_0 = x]$

Figure 5.6 clarifies the problem of linear extrapolation in view of the quantities we are going to solve: $V(t,x) = \mathbb{E}[\Phi(.)|X_t = x]$.

We assume an upper bound u and the underlying Markovian Ito Process X_t to be normally distributed - here displayed as a binomial tree (as explained in Section 1.2).

We would choose the domain boundary such that we have for the process at u: $\mathbb{P}(X_t \leq u) = \epsilon$ with ϵ sufficiently small - i.e. the probability of the process to reach or go beyond this point u is supposed to be close to zero. Otherwise we would cutoff an amount which influences the behavior of the numerical solution and underestimates the conditional expectation at value time $V(0,x)$.

An Appropriate Space Domain for the Valuation PDE 4.2

We would like to propose a procedure which lets us apply linear extrapolation (5.22) at space boundaries, where the probability that the state variable will reach this point is close to zero. Recall our Markovian System (4.4) represented by the dynamics of (X_t, Y_t) in the case of constant parameters. We usually would like to know the expected value of a future payoff to the valuation time $t = 0$ at system state $X_0 = 0$ and $Y_0 = 0$, which corresponds to the actual level of the yield curve.

$\diamond$ Firstly we have the domain boundaries for X_t: For $Y_t = y$ fixed we can analytically solve the resulting one-dimensional SDE like we did in (1.3) and obtain a gaussian distribution for X_t:

$$X_T \sim \mathcal{N}\left(\frac{y}{\kappa}(1 - e^{-\kappa T}), \frac{\eta^2}{2\kappa}(1 - e^{-2\kappa T})\right)$$

where in the case of maximized variance with $\kappa = 0$ and $y = 0$ (see Section 1.5) we would have:

$$X_T \sim \mathcal{N}(0, \eta^2 T)$$

To truncate only a small amount with $\mathbb{P}(X_t \leq u) = \epsilon$ we usually take the interval of $+/-5$ standard deviations.

$\diamond$ For the second state variable Y_t in the case of constant parameters we can calculate its exact value in Section 4.4

$$Y_T = \frac{\eta^2}{2\kappa}(1 - e^{-2\kappa T})$$

where in the case of $\kappa = 0$ we would have: $Y_T = \eta^2 T$.

Proposition 5.11. *An appropriate space domain in combination with linear extrapolation for the valuation PDE (4.2) with constant parameters is therefore found at:*

$$\Omega = [-5\sqrt{V}, +5\sqrt{V}] \times [0, V] \quad with \quad V := \frac{\eta^2}{2\kappa}(1 - e^{-2\kappa T})$$

Respectively when $\kappa = 0$:

$$\Omega = [-5\eta\sqrt{T}, +5\eta\sqrt{T}] \times [0, \eta^2 T]$$

We presented here the case of constant parameters. For more general specifications of the volatility parameter η_t - as introduced in Section 6.2 - we would choose the more general domain:

$$\Omega = [-5\sqrt{V}, +5\sqrt{V}] \times [0, V] \quad \text{with} \quad V := \int_0^T \eta_\tau^2 d\tau$$

The Error of Linear Extrapolation at the Boundaries

Here we like to show the effect when truncating the domain to early, i.e. applying the linear extrapolation at domain boundaries which are chosen smaller than proposed in Proposition 5.11. We tested the effect of choosing several boundaries for the value of a 5-year zero-coupon bond valued with the Valuation PDE (4.2) - i.e. two space dimensions.

Scheme	**Domain**	$[\text{lb}, \text{ub}] \times [0, 0.05]$
	Grid	chosen such: $\Delta_t = \Delta_x = \Delta_y = 0.01$
	Discretization	$\mathcal{D}^3$ (first) + Upwind (second)
Model	η	0.1
	κ	0.1
Product	ZCB, Maturity	5.0

The following table shows the numerical bond prices compared to the analytical bond value (4.10) at state $X_0 = Y_0 = 0$ and with several space intervals in the first space dimension.

Interval	Bond Price	Error
Analytical	0.818067595	–
$-0.88 \leq x \leq 0.88$	0.81808707	2E-05
$-0.3 \leq x \leq 0.3$	0.81535133	3E-03
$-0.2 \leq x \leq 0.2$	0.80359859	1E-02
$-0.1 \leq x \leq 0.1$	0.75886563	5E-02
$-0.05 \leq x \leq 0.05$	0.72170342	1E-01

Reviewing Proposition 5.11 we find that the domain boundary will have to be chosen larger, when the volatility parameter η increases.

We observe that choosing the correct domain boundaries calculated according to Proposition 5.11 the scheme reproduces the exact bond price (4.10): In the second row of the above table the error (2E-05) corresponds to the one we would expect when choosing a second order discretization and weighting $\theta = 0.5$, i.e. we have second order accuracy (see upper figure on the next page).

Truncating the domain to early (lower figure on the next page) introduces an error which becomes convected into the entire domain and even influences the market bond price at $(X_t, Y_t) = (0, 0)$. The truncation error would increase, if we chose an increased volatility parameter η.

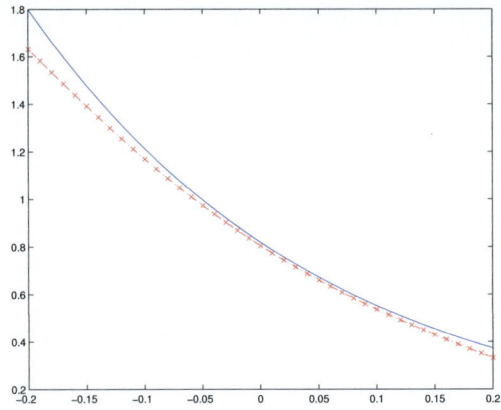

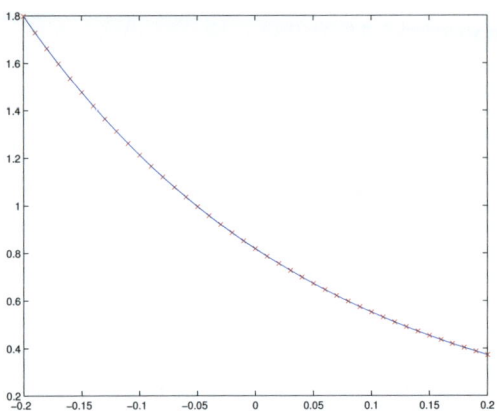

Figure 5.7.: Effect of Wrong Boundaries: First Figure with $[-0.2, 0.2]$
and Second Figure with $[-0.88, 0.88]$ interval for the first
state variable

5.6. Digression: Working on Nonuniform Grids

Valuation of derivative products through PDE's with their corresponding payoff profiles is characterized by crucial regions of sharp gradients and even discontinuous points in the final-time condition $\Phi(.)$ - usually at the level of the strike price.

On the other hand in regions far away from the strike level the solution has a very inconspicuous behavior which is often only of secondary relevance. It is therefore useful to efficiently place the number of spatial grid-points which one is equipped on.

In particular we would like to have a very dense grid in regions with substantial change and where the probability is high that the underlying falls in this region at payoff time.

In addition we can make use of the advantage that for product valuation the point of a sharp gradient or discontinuity is usually known in advance through the specification of the product to be valued. As we did work until now only with symmetric finite differences, we firstly have to give the more general finite differences on non-uniform grids where $\Delta_{m+1} \neq \Delta_m$.

Unsymmetric Finite Differences

The following finite difference formulas are obtained through Taylor expansion in the same manner as in Section 5.1. We do this explicitly for the first derivative to show, that we are left with first order truncation error. We now have with $\Delta_{m+1} = x_{m+1} - x_m$ and $\Delta_m = x_m - x_{m-1}$:

$$U_{m-1} = U_m - \partial_x u \Delta_m + \frac{1}{2} \partial_{xx} u \Delta_m^2 + O(\Delta_m^3)$$

$$U_{m+1} = U_m + \partial_x u \Delta_{m+1} + \frac{1}{2} \partial_{xx} \Delta_{m+1}^2 + O(\Delta_{m+1}^3)$$

Subtraction such that we can add the first order derivatives we obtain:

$$\frac{\partial u}{\partial x} = \frac{[U_{m+1} - U_m]}{2\Delta_{m+1}} - \frac{[U_{m-1} - U_m]}{2\Delta_m} + O(\Delta_{m+1}) + O(\Delta_m)$$

For the second derivative we obtain similarly:

$$\frac{\partial^2 u}{\partial x^2} = \frac{2}{\Delta_{m+1} + \Delta_m} \left[\frac{U_{m+1} - U_m}{\Delta_{m+1}} - \frac{U_m - U_{m-1}}{\Delta_m} \right]$$

Grid Transformation

We follow here the analysis of [23, Chapter 5]. There are actually two ways how to apply an coordinate transformation $x \mapsto T(x)$ to a finite difference scheme for a PDE defined on an uniform Grid G.

⋄ Using the Jacobian $J(x) := \frac{dT(x)}{dx}$ we can calculate via the chain rule, what our PDE becomes in terms of this coordinate transformation. Since this is not always easy to resolve, we do not calculate this exactly but can write the difference scheme equation under the use of the discretized version of $\bar{J}(x)$. For further details see [23, pp. 157]. This procedure allows us to let the uniform grid G stay fixed.

⋄ The other approach is to solve the finite difference scheme for the original PDE on the transformed grid $T(G)$ under the use of unsymmetric finite differences from above.

The here reviewed grid transformation - following [23, pp.167] - has the advantage that we can compact our grid at one arbitrary point K on our chosen domain. The grid transformation maps the unit interval on a desired domain $[l, u] \subset R$:

$$T : [0, 1] \to [l, u]$$
$$x \mapsto K + \alpha \sinh \left[c_2 x + c_1[1 - x] \right] \tag{5.26}$$

Where we have[12]:

$$c_1 := \text{arsinh}\left(\frac{l - K}{\alpha} \right) \quad \text{and} \quad c_2 := \text{arsinh}\left(\frac{u - K}{\alpha} \right)$$

The Jacobian writes:

$$J(x) := \frac{dT(x)}{dx} = b\sqrt{\alpha^2 + [T(x) - K]^2} \quad \text{with} \quad b \in \mathbb{R}$$

As we can see $J(x)$ becomes minimized - correspondingly the size between two points is smallest - exactly at the point K where a high density is desired. The parameter α determines the density around the dense point K.

---

[12]$\sinh(x) := \frac{1}{2}(e^x - e^{-x})$ and its inverse $\text{arsinh}(x) := \log \left[x + \sqrt{x^2 + 1} \right]$

The first figure shows how the grid transformation (5.26) maps a uniform grid on the x-axis to a compacted grid on the y-axis. The second figure shows a two-dimensional nonuniform grid with one dense point.

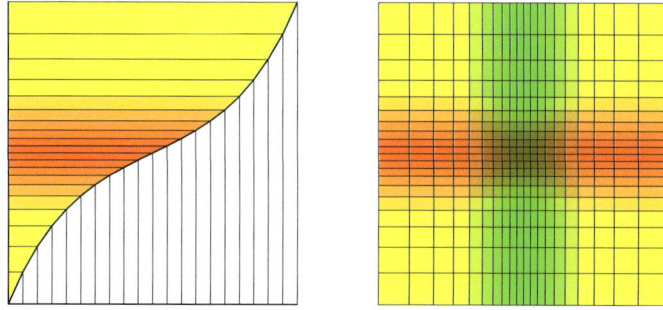

Figure 5.8.: The Grid Transformation (left) and a dense grid (right)

6. Practical Considerations

In this chapter we would like to review of the Markovian model setup (4.4) from a practical perspective and emphasize some advantages compared to other short-rate models (Table 3.1) and its behavior in view of whole yield models - like the Libor Market Model. We consider the following four main issues in detail.

$\diamond$ *Markovian Structure.* Due to the model's dependence on the particular current state we were enabled to derive a valuation PDE which in turn enables us - in combination with our robust finite difference schemes from Chapter 5 - to apply a fast deterministic pricing to any future payoff profile emerging from a interest rate product or derivative claim. Furthermore we are able to apply the *Bellman's Backward Induction Algorithm* to products equipped with an early-exercise feature, which we shortly explain in Section 6.1.

$\diamond$ *Flexible Volatility Specification.* Reviewing the model from (4.4) we have seen that it allows to arbitrarily - even stochastically perturbed - specify the short-rate volatility parameter η, only under the restriction to depend on the current system state. How and what kind of specifications are used to make the Markovian Model consistent with e.g. the Libor Market Model is reviewed in Section 6.2.

$\diamond$ *True Stochastic Volatility.* We pointed out in Remark 4.3 that the bond price does not depend on the short-rate volatility. That this feature gives a more realistic model setup compared to other short-rate models when introducing the volatility to be itself a stochastic process is explained in Section 6.3.

◇ *Exogenous Yield Curve Model.* Like the HJM and Libor Market
Model the model does have an exogenous model character since
the initial market yield curve $f^M(0,t)$ serves as a direct model input into the reaction term of the valuation PDE (4.13). In contrast
to traditional short-rate models we do not have to *calibrate* the
model - i.e. adjust the model parameters - to match these data.
However, in view of calibrating the model to the plain derivatives
market we have to test it on its flexibility to reproduce observed
market phenomena such as volatility smiles.

6.1. Early Exercise Products and Optimal Control Problems

Exotic derivative payoffs - like the Bermudan Swaption discussed in Section 2.3 - are usually equipped with an early exercise feature.

The valuation of those product types in combination with the current underlying dynamical structure can be treated very efficiently in a more abstract mathematical setting using well-known optimization techniques, as they belong to the class of *Optimal Stopping Problems*, which again are a subclass of *Stochastic Optimal Control Problems*.

We do not delve into the theory and formalism of stochastic optimal control problems, since this would go beyond the scope of this book - see [21, pp.42] for further details.

Instead we would like to emphasize in a more informal way for the specific product of a Bermudan Swaption, how we can make use of the Markovian Structure to combine the valuation PDE represented by a suited finite difference scheme from Chapter 5 with a Backward Optimization Algorithm.

The Optimality Principle ...

In view of solving an optimal stopping problem the Markovian Structure of the underlying stochastic dynamics are again advantageous for us, since their dependence on the particular current system state let us apply the *Optimality Principle* This principle states that the optimal decisions, which will be made for the period $T_i \ldots T_n$, are independent of the decisions which have been made at the period $T_0 \ldots T_{i-1}$. That in turn enables us to apply Bellmann's Backward Induction Algorithm in combination with the numerical finite difference scheme.

... applied to the valuation of the Bermudan Swaption

Review the Bermudan Swaption Product from Section 2.2: The holder of such a product form has the right to enter in an Interest Rate Swap at some previously determined exercise times $\mathcal{T} = \{T_0, \ldots, T_n\}$. At each of those fixed dates the holder has to decide whether to enter into the underlying Swap Structure or hold the Bermudan Swaption until the next exercise time. We already noted that this gives us an optimization problem at each of those stopping times. Given a current system state from the model setup (4.1) i.e. the filtration $\mathcal{F}_{T_i} = (X_{T_i}, Y_{T_i})$ we have to solve at each $T_i \in \mathcal{T}$ for each particular outcome (x, y):

$$V_{\text{BSwpt}}(T_i; x, y) = \max \left[\underbrace{V_{\text{Swap}}(T_i; T_i)}_{\text{Exercise}}, \underbrace{V_{\text{BSwpt}}(T_i; T_{i+1})}_{\text{Hold}} \right]$$

$$= \max \left[V_{\text{Swap}}(T_i; T_i; x, y), \right.$$

$$\left. \mathbb{E}^{RN} \left[D(T_i, T_{i+1}) V_{\text{BSwpt}}(T_{i+1}; T_{i+1}) \big| X_{T_i} = x, Y_{T_i} = y \right] \right]$$

The Backward Optimization Algorithm recursively maximizes the value at each time T_i and particular space state (x, y) over all stopping times $T_i \in \mathcal{T}$ starting at T_n.

How this can be combined with risk-neutral valuation in terms of the valuation PDE respectively the finite difference scheme is summarized in the following algorithm:

Algorithm 6.1. *Given a Bermudan Swaption with Tenor* $\mathcal{T} = \{T_0, \ldots, T_N\}$

```
For Each T_i on Time Scale
  1. Set the Final-Time to Tenor Date T_i
  2. Rollback the Difference Scheme to previous Tenor Date T_i-1
  3. SwaptionValues <- max(SwapValues, SwaptionValues)
Next T_i-1
```

In this context we also refer to the C++ valuation algorithm in Section 7.3.

6.2. Local and Stochastic Volatility Specifications

In Section 4.4 we analyzed the model with constant parameters. Especially with a constant short rate volatility parameter η we have shown that the model setup (4.4) incorporates already known short rate models - in particular the Hull-White Model and the Jamshidian Model. The essential advantage to work with the more general model specification as presented in Theorem 4.1 is that we have the flexibility to choose the short-rate volatility parameter $\eta(t, x, y)$ to depend on the particular current system state (x, y) and furthermore we are able it to let it be driven itself by an additional Markovian Process.

A very short note on the Concept of Local Volatility

Engineered by B. Dupire in 1994[1] and firstly applied to the equity derivatives market, the local volatility model has turned out to be a generalized Black & Sholes Model, which automatically fits the entire Plain Vanilla Options Market.

To achieve this and since the only free parameter is the volatility parameter, it becomes defined to be a function of strike K and maturity T - which in turn can be interpreted as future spot and time points:

$$\sigma_{\mathrm{LV}} := \sigma_{\mathrm{LV}}(K, T)$$

Since now the volatility coefficient is time and space dependent, our model dynamics - given in some similar manner as in (1.21) - now become *non-homogenous*. Now combining the knowledge of the underlying model probability densities, which are basically determined by the Brownian Motion part of the underlying process through (1.7), we can fit those densities to the market implied ones, which can be backed out from the market's vanilla call option prices C:

$$p^{\mathrm{Market}}(t, S, K, T) = \frac{\partial^2 C(S, t; K, T)}{\partial K^2}$$

Comparing those densities with the forward PDE dynamics of the model implied density (similar to (1.8)) one is then able to back out the volatility coefficient, which then yields the famous Dupire formula.

---

[1]Pricing with a smile, *RISK*, 7(1):18-20, January 1994

The right interpretation of local variance is given as the expectation of the process' future spot variance σ_T^2, conditioned on the underlying price to be equal to K at time T.

$$\sigma_{\mathrm{LV}}^2(K, T) = \mathbb{E}[\sigma_T^2 | S_T = K]$$

For a derivation of this result, see [24, pp.13]. Modulo the assumption of the underlying process structure (Diffusion Process) this would be inferred as a *non-parametric calibration approach*.

Beyond the original Dupire approach, several approaches - especially in the interest rate world (e.g. [15]) - try to find or assume a *Functional Form* for the above conditional expectation and do a *Parametric Calibration* to calibrate the parameters such that the model reproduces the observed plain vanilla prices.

State Dependent Short-Rate Volatility

Many of the now introduced concepts like *Displaced Diffusion* and *Constant Elasticity of Variance* where the volatility parameter depends on the *current* system state (t, x, y) have been previously applied to the Libor Market Model (Section 3.3). These also called *Local Volatility* specifications have shown to be very efficient in view of calibrating the model to market data in the sense of Section 2.3. For a detailed overview of state dependent volatility specifications in the Libor Market Model, we refer the reader to [25, Chapter 4]. Here we give an overview of how the parameter $\eta(t, x, y)$ has been specified in the model setup (4.4) recently to efficiently calibrate this model to observed market data.

- ⋄ A *Time Dependent* Parameter $\eta(t)$ would be the first extension of the case of constant parameters. This would yield similar one-dimensional model dynamics as in Section 4.4 as we would still be able to solve the resulting ODE for Y.

- ⋄ *State Dependent* - in particular short-rate dependent - volatility was introduced in [16] for the one-factor case with the following specification:

$$\boxed{\eta(t, x, y) := \sigma r_t(x)^\gamma}$$

  where $\sigma = $ const and the short rate r_t defined as in (4.8). Therefore this extended model can be calibrated under the use of three constant parameters κ, σ and γ.

⋄ A *Constant Elasticity of Variance* (CEV) approach has been discussed in [17]:

$$\boxed{\eta(t, x, y) := \sigma(t) R(t, x, y)^{\alpha}} \tag{6.1}$$

with a time-dependent parameter $\sigma(t) \in \mathbb{R}$ and $\alpha \in \mathbb{R}$. $R(t, x, y)$ denotes a *variable* for an abstract interest rate quantity which becomes determined by the specific market segment the model is supposed to be calibrated to. For instance we could specify $R(t, x, y)$ to be the short rate $R(t, x, y) := r(t, x, y)$, a term rate $R(t, x, y) := F(t, T_1, T_2; x, y)$ or a Swap Rate $R(t, x, y) := S_{a,b}(t, x, y)$. Constant Elasticity in the present meaning says, that a one percentage increase in the interest quantity R implies a constant α percentage increase in the volatility:

$$\frac{d\eta}{dR} \frac{R}{\eta} = \alpha$$

⋄ *Displaced Diffusion* is usually viewed as a model with a mixture of normal and log-normal behavior. Review the two SDE's in Section 1.4, which we solved analytically and showed that their solutions are lognormal respectively Gaussian distributed. This was mainly influenced by the specification of the volatility parameter: Choosing it to depend explicitly on the particular system state yields lognormal behavior (1.21) where independence yields Gaussian statistics (1.24). Now introducing a time-dependent parameter $m(t) \in [0, 1]$, which mixes those both approaches we obtain with $R(t, x, y)$ being again a variable for an interest-rate quantity:

$$\boxed{\eta(t, x, y) := \sigma(t) \big[m(t) R(t, x, y) + \big(1 - m(t) \big) R_0 \big]} \tag{6.2}$$

where R_0 is constant - in particular the current market rate to be modeled. We see that choosing $m(t) \equiv 1$ yields lognormal behavior of the resulting process whereas $m(t) \equiv 0$ yields gaussian statistics. For $0 < m(t) < 1$ we have a mixture of both distributions.

Adding Stochastic Volatility

Stochastic Volatility Models i.e. letting the volatility parameter be driven by an additional stochastic process have their historical roots with the *Heston Model*[2] firstly used only in the equity world. This approach has been also recently transferred and applied to the Libor Market Model - for a detailed review see [25, Chapter 6].

Our Markovian Yield Curve model in combination with stochastic volatility was firstly analyzed in [26] and more recently in a multi-factor setting in [2]. This is surely possible in the model environment of Theorem 4.1 as we are allowed to let the parameter η to dependent on the system state (x, y) and in addition to let it be driven by an additional Markovian Process (see also the discussion in Section 4.1). We do this by introducing the so-called *Square-Root* Process Z_t defined by:

$$dZ_t = \beta(t)\big[1 - Z_t\big]dt + \epsilon(t)\sqrt{Z_t}dW_t \tag{6.3}$$

with long-term mean parameter $\beta(t)$ and a *Volatility of Variance* parameter $\epsilon(t)$. The positive square-root of this additional process then becomes directly integrated into the Displaced Diffusion Specification, so we have now at a system state (t, x, y, z):

$$\boxed{\eta(t, x, y, \mathbf{z}) := \sqrt{z}\sigma(t)\big[m(t)R(t, x) + (1 - m(t))R_0\big]} \tag{6.4}$$

Consistency with the Libor Market Model

As the Libor Market Model (Section 3.3) is still the main reference of interest rate models, there is even a way to derive a more specific volatility structure with $R(t, x, y) := f(t, T; x, y)$ with the aim to get the Markovian model (4.4) consistent with that market model. This has been proposed in [2].

In terms of forward rates $f(t, T)$ we define its dynamics in terms of displaced diffusion and stochastic volatility:

$$df(t, T) = \sqrt{z}\lambda(t)\big[m(t)f(t, T) + (1 - m(t))f(0, T)\big]dW^{SL} + \mu dt$$

---

[2]S. Heston, A closed-form solution for options with stochastic volatility with applications to bond and currency options, Rev. Financial Stud., 6 (1993), pp. 327–343.

For our Markovian Model we have from Proposition 4.4:

$$df(t,T) = \frac{\partial G(t,T)}{\partial T} \eta dW^{RN} + \mu dt$$

We are not interested in the drift terms so we simply write μdt, but we would like to transfer the volatility specification from the LMM to our Model. This is now simply achieved through a parameter comparison. Note that although we are working under different measures (Spot-Libor vs. Risk-Neutral) we can see from (3.14) that those both do not differ much in this here considered maturity-continuous version. Beyond that even if we transfer both model to one unit measure, this measure change will not affect the volatility coefficient according to Remark 1.13. Parameter Comparison and solving for η yields the following *Forward Rate Volatility Specification*:

$$\eta(t,x,y,\mathbf{z}) := \left[\frac{\partial G(t,T)}{\partial T}\right]^{-1} \sqrt{z_t}\sigma(t)\big[m(t)f(t,T) + (1 - m(t))f(0,T)\big]$$

How this procedure works for swap rates - which is in turn not that straightforward to do, we refer to [2].

The Extended Valuation PDE

With the Square-Root Process (6.3) we have included an additional Markovian Process into the Markovian System (4.4). In comparison to Theorem 4.1 we have an extended three-dimensional SDE system, which we present here at a glance in combination with the volatility parameter specified as in (6.4):

$$dX_t = \big[Y_t - \kappa(t)X_t\big]dt + \eta_t dW_t^{RN}$$
$$dY_t = \big[\eta_t^2 - 2\kappa(t)Y_t\big]dt$$
$$dZ_t = \beta(t)\big[1 - Z_t\big]dt + \epsilon(t)\sqrt{Z_t}dW_t^{RN}$$

In view of the parameters we now have $\eta_t := \eta(t,x,y,z)$ exogenously stochastically driven and specified from above, κ is usually chosen to be constant and the deterministic parameters $\beta(t)$ and $\epsilon(t)$ time-dependent. Compared to the two-dimensional valuation PDE from Theorem 4.2 we

now have an extended valuation PDE with three spatial dimensions
for the risk-neutral expectations (4.12) as a direct application of the
Feynman-Kac Theorem 1.4:

$$
\underbrace{[f(0,t) + x]}_{=r(t,x)} V = \frac{\partial V}{\partial t} + [y - \kappa x]\frac{\partial V}{\partial x} + \frac{1}{2}\eta_t^2\frac{\partial^2 V}{\partial x^2}
$$

$$
+ [\eta_t^2 - 2\kappa y]\frac{\partial V}{\partial y}
$$

$$
+ \beta(t)[1 - z]\frac{\partial V}{\partial z} + \frac{1}{2}\epsilon(t)^2\frac{\partial^2 V}{\partial z^2} \tag{6.5}
$$

In view of the numerical solution we are most readily equipped with an
efficient ADI solver from the three-dimensional Craig & Sneyd Scheme
(5.13) which is suitable to cope with multiple spatial dimensions.

The Valuation Models Reviewed

In conclusion considering the model SDE's from Chapter 4 and the
extended version from the current chapter we now have *three* versions of
the valuation PDE, which we are all be able to solve with our according
dimension specific ADI Scheme (5.13):

◇ In the case of constant parameters we have the one-dimensional
   SDE (4.14) and the resulting PDE (4.16) with one spatial dimension - which corresponds to a valuation in the Hull-White Model
   setting.

◇ In the more general two-dimensional model setup (4.4) we have the
   two-dimensional valuation PDE (4.13) with the flexibility of more
   general state dependent deterministic volatility specifications from
   above.

◇ The extended version with stochastic volatility introduces the
   three-dimensional PDE (6.5) from above.

6.3. True Stochastic Volatility

In the preceding section we have shown that the Markovian Yield Curve Model is useful for practical reasons since we can even let the volatility be driven by an additional stochastic process. The reader might wonder, if this flexibility - i.e. incorporating stochastic volatility into the model setup - might also be possible in a short-rate model context. So we could ask why not simply extending the models from Table 3.1 in the sense that we assume:

$$dr_t = [b - ar]dt + \sigma_r(r_t, \mathbf{Z_t})dW$$

with Z_t defined as in (6.3).

However it has turned out that this additional *source of risk*, which we are including through the assumption of stochastic volatility, falls short of a realistic modeling environment. The value $V(r_t, Z_t)$ of a derivative product such as a caplet should be sensitive to movements in the volatility parameter σ.

$$\frac{\partial V(r, Z)}{\partial \sigma_r} \neq 0$$

This inherent risk is tried to be modeled by introduction of stochastic volatility, i.e. including an additional source of noise being represented by a second Brownian Motion.

The following observation has now been made in view of bond prices, whose model value in terms of the short rate is reviewed here from (3.2):

$$P(t, T; r_t) = A(t, T) \exp[-B(t, T)r_t]$$

Where $A(t, T)$ and $B(t, T)$ can be found in the given references. It is important to realize that they usually do depend on the volatility, so we have $A(t, T, \sigma)$ and $B(t, T, \sigma)$. This in turn implies that the model bond price does also depend explicitly on the short rate volatility σ_r. So we have for the standard short-rate models:

$$P(t, T) = P(t, T; \mathbf{r_t}, \sigma_r) = P(t, T; \mathbf{r_t}, \mathbf{Z_t})$$

As the bond price is itself sensitive to the short rate and the volatility:

$$\frac{\partial P(t, T_1)}{\partial r} \neq 0 \quad \text{as well as} \quad \frac{\partial P(t, T_2)}{\partial z} \neq 0$$

we are now able *replicate* any change in a derivative value V - which we explained in Section 1.8 by setting up a suitable portfolio strategy (ϕ_t, ψ_t) against it. Given the system state (r, z) we firstly can write in the sense of Ito:

$$dV(r, z) = \frac{\partial V}{\partial r} dr_t + \frac{\partial V}{\partial z} dZ_t + O(dt)$$

As the bond price is sensitive to both quantities r_t and Z_t we are furthermore able to set up a replicating portfolio (ϕ_t, ψ_t) in the sense of (1.43) consisting of two bonds with different maturities $P(t, T_1)$ and $P(t, T_2)$.

$$dV(r, z) = \phi_t \frac{\partial P(t, T_1)}{\partial r} dr_t + \psi_t \frac{\partial P(t, T_2)}{\partial z} dZ_t$$

This implies that the volatility risk inherent in a derivative can be eliminated by trading in zero-bonds $P(t, T)$ alone. This implication seems wrong on the first view and furthermore it has shown empirically it does not match market reality.

Coming back to our more abstract Markovian Representation we have seen in Remark 4.3, that the Model Bond Price $P(t, T; x, y, z)$ from (4.10) is *independent* of the volatility η. Given a system state (x, y, z) for bond price we have here:

$$\frac{\partial P(t, T; x, y, z)}{\partial z} = 0$$

Therefore that model models stochastic volatility *truly* in correspondence with market reality as we are not able to eliminate volatility risk by trading in zero-coupon bonds. Because of this the model is out of the class of *True Stochastic Volatility Models*.

6.4. Calibration to Market Data

The main emphasis on this thesis was to investigate and implement product valuation by PDE methods in the presence of Markovian Yield Curve Dynamics. Most important from the practical perspective is how well the model parameters can be calibrated to fit observed market data and furthermore observed market phenomena.

The market data to which the model has to be calibrated to can be grouped by:

⋄ The Market Yield Curve $f^M(0, T)$

⋄ Cap Volatilities

⋄ Swaption Volatilies

Considering the yield curve data $f^M(0, T)$ we do not have to calibrate model parameters since they serve as an exogenous model input.

In view of the market of derivative products their prices are quoted by *Implied Volatilities*, which are derived by an inversion of Black's Formula.

Considering the different model versions we have to calibrate their parameters such that they reproduce these volatilities and through that the market prices of the standard interest rate derivatives market.

The highest calibration flexibility is surely achieved with the extended valuation PDE (6.5), as we have here the largest number of parameters, which in turn can be chosen to be time-dependent. We tested this model quality in Section 6.5. For further details on the calibration issue we refer to [17] and [26] and [2].

6.5. Model Implied Volatility

We remarked in Section 6.4 that a valuation model should be flexible enough to match observed market phenomena. The most important issue when investigating a interest rate model is, how well it matches *Implied Volatility Surfaces* in the derivatives market, i.e. the volatility parameter as a function depending on strike and maturity:

$$(K, T) \mapsto \sigma(K, T)$$

The market uses Black's Model (2.33) to price plain vanilla derivative products such as caps and european swaptions. As the only free parameter is the forward rate volatility and furthermore the valuation formula is monotonic in this parameter, derivative prices get quoted by inverting Black's Formula which yields the implied volatility.

This quantity is usually not constant as assumed in the original model but varies with strike and maturity of the corresponding derivative: $\sigma = \sigma(K, T)$. What is usually observed is that such a volatility surface has kind of a convex shape with a minimum around the strike level - for a fixed maturity this phenomena is called the *volatility smile*. Furthermore it has been observed that this volatility surface can also be *humped* or *skewed*. For further details on implied volatility in the interest rate derivative market we refer the reader to [9, pp.78].

As we have to calibrate the model parameters of a valuation model such that this model reproduces the prices for plain derivative products, the model should have the flexibility to match those observed volatility surface structures (smiles, skews, humps).

We have tested here the *Model Implied Volatiliy* for a caplet under the use of:

    ◇ The two-dimensional valuation model in combination with displaced diffusion specification (6.2) of the short rate volatility parameter

    ◇ The extended valuation model (6.5) with displaced diffusion.

Model implied volatility enables us to see what kind of shapes of implied volatility these models are capable to generate.

This has been archived by the use of the following algorithm:

Algorithm 6.2. *Model Implied Volatility for a Caplet for a fixed Maturity:* $K \mapsto \sigma_T(K)$

```
For Each K Do
  1. Calculate the Model PDE Price for Strike K
  2. Invert Black's Formula: Get Model Implied Volatility
Next K
```

Valuation PDE 6.5 with Displaced Diffusion 6.2

The following table lists the parameters we used for the following figure. In the context of displaced diffusion, the parameter $m = 1.0$ corresponds to an underlying lognormal distribution whereas $m = 0.0$ would be a normal distribution - see also Section 6.2. We observed that the valuation PDE (4.13) is able to produce flat or skewed volatility curves whereas using the extended version with stochastic volatility we are able to obtain convex (smile) shaped curves. As we already mentioned we now see that the extended model is more flexible in the sense of that it is capable of matching more complex shapes of volatility curves.

m	1.0 versus 0.0
σ	0.2
κ	0.02
Strike	0.04135 $[X_0 = 0]$
Maturity	1.0
Tenor	1.0

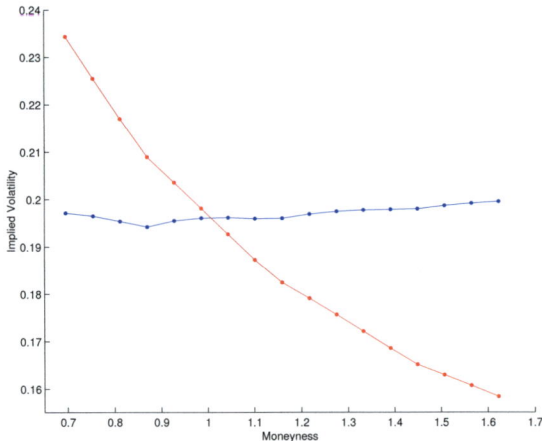

Figure 6.1.: Valuation PDE (4.13): Model Implied Volatility is flat for $m = 0$ and skewed for $m = 1$

Extended Valuation PDE 6.5 with Displaced Diffusion 6.2

The extended valuation model is not only capable to produce skews but to generate convincing implied volatility smiles, which makes it preferable compared to the two-dimensional valuation model (4.13), since it is able to fit diverse observed market smiles.

m	1.0 versus 0.0
σ	0.2
β	0.2
ϵ	0.85
κ	0.02

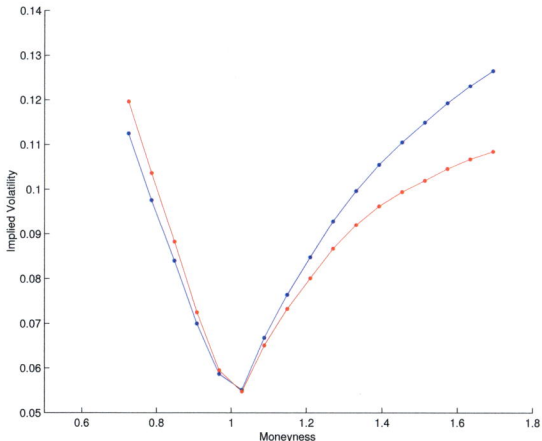

Figure 6.2.: Extended Model (6.5): The Model's Implied Volatility Smile for $m = 0$ and additionally skewed for $m = 1$

7. Design Issues and C++ Implementation

In this chapter we like to give an overview of an implementation approach, which can be separated into the **Back End** with the Finite Difference Solver represented by the class FDScheme, which is based on the ADI Method in particular the Scheme (5.14). Secondly we have a **Front End** with the class PDEProduct, through which the product valuation is done in terms of our valuation PDE's. An intense use of polymorphism - represented by several interfaces (respectively abstract base classes) enables the programmer to achieve high flexibility in terms of dimension independence - in particular in view of valuation models with varying dimensions, discretization schemes, value management and product functionality.

In the following we do not discuss every single method in detail. We would rather like to give an overview of core classes and methods to give the reader an intuition how they can be used to efficiently value different interest rate products under the use of the valuation PDEs from Chapters 4 and 6.

7.1. Components of the Finite Difference Scheme

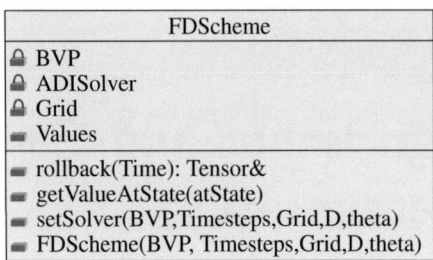

Figure 7.1.: Class Diagram for the Finite Difference Scheme

The back end class which is used by the product valuation later on consists of a pointer on a Boundary Value Problem (Class BVP), a pointer on a Value Object (Class ValueTensor), a pointer on a Grid Object (Class Grid) and a pointer on an Solver Object ADISolver. A scheme is instanced by a BVP, a space grid, the number of time steps and a vector with codes for the particular applied discretization methods. With those as arguments the corresponding dimension-specific ADI scheme solver is instanced and the dimension of the value tensor values is specified. Exemplarily when having a valuation model as a BVP, 100 time-steps, a grid object, the discretization vector and the implicitness of the scheme specified we are able to instance a weighted ($\theta = 0.5$) scheme with the constructor:

FDScheme scheme(model,100,grid,discretization,0.5);

When doing a particular valuation we then call the rollback-method which solves the scheme backwards over time and returns a tensor object with the calculated values at a specified value time. Theses values are also stored in the tensor object values where they can be accessed and modified at any time.

We explain now in more detail the particular components used to instance a scheme by the above constructor.

ADI Scheme and Solvers

We start explaining the abstract class `ADIScheme` which itself is dimensionless and only responsible for the specific scheme procedure - i.e. it generates the linear equation system which is solved at every particular grid point according to the specific ADI scheme equation (5.13). We now describe two main methods which are used to rollback the scheme in time. The grid management is sourced out into a separate `Grid`-Class which will be discussed below.

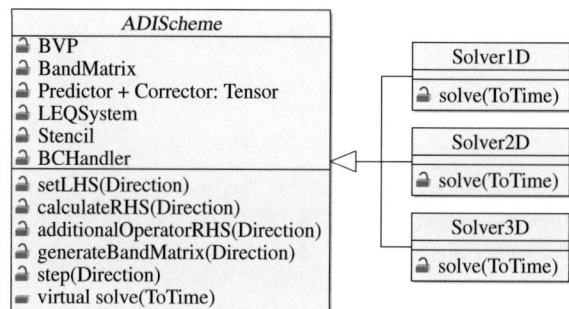

Figure 7.2.: UML Diagram for the ADI Scheme

We discuss here the `solve` and `step` Method in more detail. The `solve`-Method is virtual and is implemented by the dimension-specific descendants (`1DSolver` - `3DSolver`). These especially manage the splitting procedure over the time grid according to (5.13). Exemplarily consider here the `solve`-Method of a two dimensional domain.

In the sense of the ADI method for every grid point of the inactive operator a linear equation system is solved and the values are inserted into the predictor value tensor for intermediate time steps respectively the corrector value tensor. In the two dimensional case this results in two main `for`-Loops for every time point, where the `step`-method gets called.

```
void ADI2D::solve(double ToTime){
 this->setToTime(ToTime);
 for(int timeIndex = 0; timeIndex < TimeSteps ; t++){
  lgs->setDimension(grid->getGridDimension(0));
  for(int yStep = 0; yStep<=grid->getDimension(1);yStep++){
   grid->setIndex(1,yStep);
   this->step(0,timeIndex);
  }
  lgs->setDimension(grid->getGridDimension(1));
  for(int xStep = 0; xStep<=grid->getDimension(0);xStep++){
   grid->setIndex(0,xStep);
   this->step(1,timeIndex+interTime);
  }
 }
}
```

The step-Method implements the specified scheme procedure (5.14) and is called at every space grid point. Firstly the respective state space and time state will be set. From that the band matrix according to the chosen difference operator is initialized. Afterwards the right hand side is built according to the ADI scheme (5.14). The same procedure applies for the left hand side of the linear equation system which then becomes solved with the chosen solver. The calculated values are either inserted into the predictor - for all intermediate scheme steps between two time points t_n and t_{n-1}- or corrector value tensor - at the last intermediate step.

```
void ADI::step(int direction, double timeIndex){
  setTimeState(timeIndex);
  generateBandMatrix(direction);
  calculateRHS(direction,this->timeState,lgs->b);
  setTimeState(t+interTime);
  generateBandMatrix(direction);
  setLHS(direction,lgs->A);
  lgs->solve();
  if(dir==bvp->getDim()-1)
    corrector->setValues(direction,grid->getIndex(),lgs->x);
  else
    predictor->setValues(direction,grid->getIndex(),lgs->x);
}
```

The Grid Class

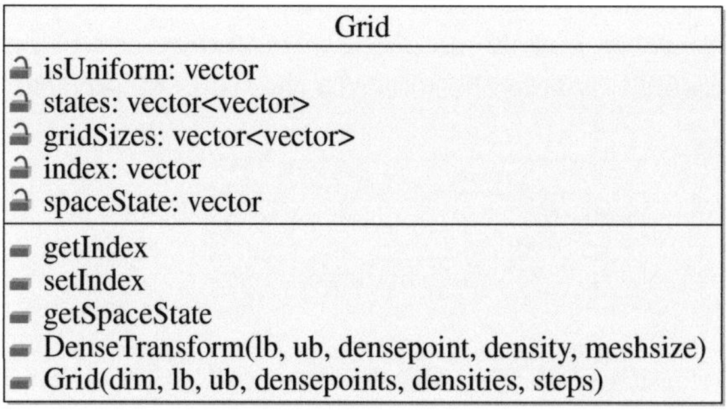

Grid
⊒ isUniform: vector
⊒ states: vector<vector>
⊒ gridSizes: vector<vector>
⊒ index: vector
⊒ spaceState: vector
▬ getIndex
▬ setIndex
▬ getSpaceState
▬ DenseTransform(lb, ub, densepoint, density, meshsize)
▬ Grid(dim, lb, ub, densepoints, densities, steps)

Figure 7.3.: The Grid Class

An equidistant grid is instanced by specifying the spatial dimension, a vector for lower and upper boundaries and the number of steps per dimension. Beyond that according to Section 5.6 it is possible to specify one point per dimension through the vector **densepoint**, at which the grid becomes compacted with the according **densities**. That functionality - in particular the coordinate transformation from Section 5.6 - is provided through the method **DenseTransform**. The calculated grid points and grid sizes are calculated in advance - i.e. at the time an object is instanced - and stored in the according vectors **spaceState** and **gridSize**.

The grid management - used mainly by the solver - is then done globally through a private integer **index** vector and its according **getIndex**-and **setIndex**-methods. At each grid point - represented through its particular integer number of grid points - the corresponding space state according to the chosen space domain is calculated and returned by the method **getSpaceState**.

```
vector<double>& getSpaceState(void) {
    for(int i=0 ; i < dimension ; i++)
        spaceState[i] = states[i][ index[i] ] ;
    return spaceState;
}
```

Boundary Conditions

A particular boundary condition is implemented by the Class `Boundary`,
where we separated out the handling of boundary conditions according
to equations (5.23) within the solver to the Class `BCHandler`. The Class

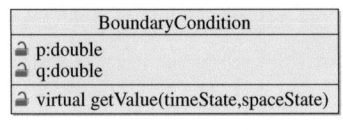

BoundaryCondition
☝ p:double
☝ q:double
☝ virtual getValue(timeState,spaceState)

BCHandler
☝ row: vector
☝ applyBoundaryConditionLHS(Boundary,[...])
☝ applyBoundaryConditionRHS(Boundary,[...])
☝ LowerBCAdditionalOperator(Boundary,[..])
☝ UpperBCAdditionalOperator(Boundary,[...])

Figure 7.4.: Classes for Boundary Specification and Handling

`Boundary` is an abstract class, where each of its descendants has to implement the method `getValue`. Its protected members `p` and `q` define,
wether we have Dirichlet ($p = 0, q = 1$), von-Neumann ($p = 1, q = 0$) or
Robbins Condition ($p \neq 0, q \neq 0$). For the numerical boundary condition (5.22) we use ($p = q = 0$).

The main task of the handler class `BCHandler`, whose methods are
called by the scheme, is to modify the first and last row of the band
matrix. The according boundary condition is recognized via the spec-
ifiers `p` and `q`. This is used together with the scheme equation to
eliminate ghost grid points from the scheme. The resulting equations
have been discussed in Section 5.5 and are directly implemented in
the methods `applyBoundaryConditionRHS/LHS`. Those when provided
with a Boundary Object (and additional arguments summed up to [...])
modify the particular scheme equation and store it in the vector `row`,
which is then further used by the solver. Furthermore - according to
the scheme equation (5.14) - as we are adding additional operators
on the right hand side in the first step - we also have to include the
according boundary conditions which is then done by the functions
`Lower/UpperBCAdditionalOperator`.

Difference Operators

The Discretization of the Differential Operators at every grid point - which corresponds to a row in the matrix of the linear system - is implemented by the abstract Class `Stencil` whose main method returns the value of the difference operator at the according space and time state. The method's arguments are the PDE Coefficients at a specific grid point. That method becomes implemented by its descendants which correspond to the particular used discretization operators. We mainly used central discretization $\mathcal{D}^3_{xx}$ (Table 5.1) for the second combined with either the Upwind (5.8), Lax-Wendroff (5.9) and the central discretization $\mathcal{D}^3_x$ for the first derivative.

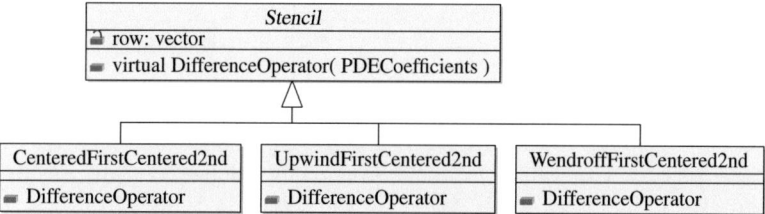

For nonuniform grids, we have the classes `NonUniformCentered` for a centered and `NonUniformUpwindFirstCentered2nd` for a combined upwind and centered discretization according to the un-symmetric discretizations, which we discussed in Section 5.6.

Linear Equation Solvers

The linear equation system solver is represented by the abstract class `LEQSystem`, where we implemented two descendants which implement the virtual method `solve`. Firstly the Thomas Algorithm - here named as TDMA (= **T**ri**D**iagonal **M**atrix **A**lgorithm) - and the standard Gauss-Seidel-Algorithms. We mainly used the TDMA solver.

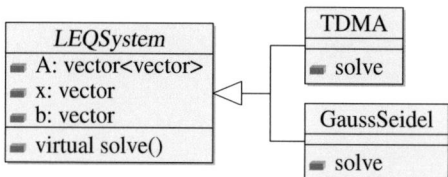

Value Management

The management of values is done in a separate abstract class `Tensor` which consists of one single array (specifically the `std::valarray`[1]) and in which the values get stored and are managed linearly. The particular space grid dimension is added on the array by the use of a `std::slice` object.

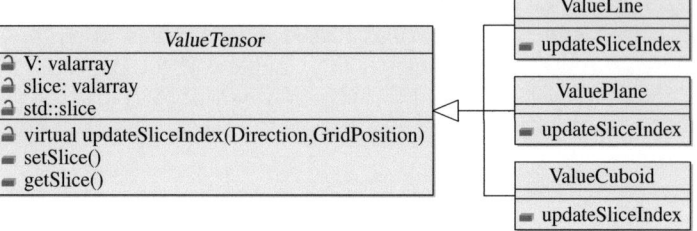

The `ADIScheme` when building a linear system - especially the right hand side - by the `step`-Method at a specific grid point does only need several single vectors with values from the previously calculated step - see first step of(5.14). This - when called by the scheme - is sliced out by the methods `updateSliceIndex`-Method and returned by the `getSlice`-Method. Since the value management in the linear array V depends on the number of spatial and the grid dimension, the slice-updating method is virtual and is implemented by the dimension specific descendants - shown here for the two-dimensional case.

```
void Plane::updateSliceIndex(int activeDirection,
                                  vector<int>& index){
  switch(activeDirection) {
    case 0:
      //Direction in second space dimension is fixed
      slice = std::slice(incr[1]*index[1],dim[0],incr[0]);
      break;
    case 1:
      //Direction in first space dimension is fixed
      slice = std::slice(index[0],dim[1],incr[1]);
      break;
  }
}
```

---
[1]std = Namespace of the C++ Standard Template Library (STL)

Adding Functionalities on the Tensor Class

We do not want to have the functionalities (mainly triggered by the demands of the product valuation) directly implemented in our value-management class `TensorValue`. Therefore we decided to use the concept of Function-Objects.

Since we have set the final-time values - equivalently to the specific product payoff - or even have to modify the calculated tensor values at several time points - triggered by the product functionality, this is done by so-called Functors or Function-Objects. These are applied on the tensor directly and modify the values corresponding to their functionality. A Functor is a class which consists of a constructor and the overloaded ()-Operator and is the resolvent in object-oriented design which replaces the Function-Pointer-Concept from C. It further depends on how many arguments are provided. Exemplarily we have a Unary Functors (which are provided with one argument) represented by the abstract base class `UFunctor`.

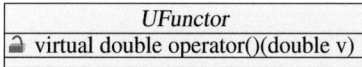

UFunctor
virtual double operator()(double v)

Also possible are e.g. Binary Functors for functions with two arguments. A Functor is applied on the Tensor-Values V by calling the following method.

```
void Tensor::apply(UFunctor& f){
    for(int i=0;i<V.size();i++) V[i] = f(V[i]);
}
```

As an example we consider here the unary Functor `Floor`, which implements the function $f(x) = \max[x - K, 0]$.

```
class Floor : public UFunctor{
  private:
    double k;
  public:
    double operator()(double v){
        return std::max<double>(v-k,0);
    }
    Floor(double k) : k(k) {}
}
```

7.2. The Valuation Model

Embedding of the Valuation PDEs

We explain now the specification of a BVP (Class `BVPSpecification`) which is part of a finite difference scheme `FDScheme` and which becomes solved by a appropriate dimension solver. The main blocks of a BVP are a pointer on a PDE object, a vector consisting of pointers to boundary objects and the final time.

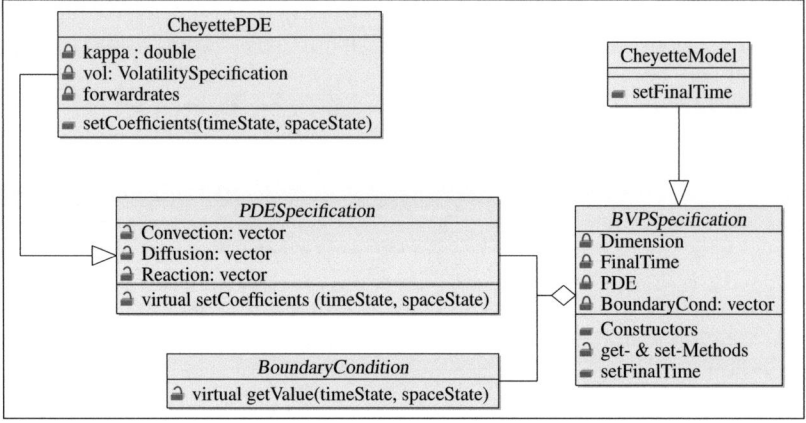

Figure 7.5.: UML Diagram for the Cheyette Model

We already explained the boundary class. The abstract class `PDESpecification` is characterized by its virtual method `setCoefficients`.

These methods get passed current grid states as arguments from which the particular PDE coefficients and boundary values are calculated and serve as input into the particular equation system.

As an example we present here the setCoefficients-Method of the two-dimensional valuation PDE (4.2) (Class Cheyette2D), which is a descendant of the class CheyettePDE. We also have the version for the one-dimensional PDE 4.16 (Class Cheyette1D) and the extended valuation PDE with stochastic volatility (6.5) (Class CheyettePDEStoVol).

```
void Cheyette2D::setCoefficients (double timeState,
                         vector<double>& spaceState) {
    r = spaceState[0] + forwardrate[timeState];
    v = vol->getValue(r,forwardrate[timeState]);
    convection[0] =  spaceState[1] - kappa*spaceState[0];
    diffusion[0]  =  0.5*v*v;
    convection[1] =  v*v - 2*kappa*spaceState[1];
    diffusion[1]  =  0;
    reaction[0]   = -r;
    reaction[1]   =  0;
}
```

The different versions of volatility specification (see Section 6.2) are implemented by a Class VolatilitySpecification and various descendants.

We work here explicitly with yield curve market data represented through a vector of forward rates $f^M(0,t)$, which are calculated out of the market's (interpolated) zero-bond curve. These data are provided in the vector forwardRates, whose dimension (in turn the refinement of the interpolation) has to match the number of time-steps we have chosen for our discretization.

The Cheyette Model as a Boundary Value Problem

The Markovian yield curve valuation models from Chapters 4 und 6 are now simply descendants from the class BVPSpecification, which consists of a version of the valuation PDE combined with numerical boundary conditions.
This derived model class furthermore overwrites the method setFinalTime, where in addition to the initialization of the final time variable the forward rates get calculated - depending on the chosen time discretization (timeSteps) - from interpolated discount factors and stored in the vector forwardrates. Those are then used directly in the Valuation PDE to calculate the specific reaction (i.e. discount) term r_t. The PDE versions with one or two spatial dimensions are instanced with:

```
CheyetteModel(int dim, double kappa,
              VolatilitySpecification* vol, int timeSteps);
```

Where for the extended PDE from Chapter 6 we use the constructor:

```
CheyetteModel(double kappa, VolatilitySpecification* vol,
              double beta, double eps, int timeSteps);
```

The Procedure in main.cpp

We explain shortly how the different components of the back end are instanced

1. Instance Volatility Specification (here CEV (6.1))

   ```
   VolatilitySpecification* Volatility = new CEV(0.2,1.0);
   ```

2. Instance a version of the Cheyette Model (here with valuation PDE(4.13)

   ```
   BVP* model =
           new CheyetteModel(2,kappa,Volatility,timesteps);
   ```

3. Instance a Grid with the appropriate space dimension, boundaries, densities, and steps

```
Grid* grid = new Grid(model->getDim(),lb,ub,
                      densepoints,densities,steps);
```

4. The model environment becomes now combined with an appropriate ADI scheme solver in an finite difference scheme object:

```
FDScheme scheme(model,timesteps,grid,discretization,0.5);
```

Now the finite difference scheme for the valuation model is set and can be used to value different products. The implementation of different product specification is discussed in the next section.

7.3. Product Valuation Routines

Product valuation is done by the use of an abstract interface consisting only of the virtual method value.

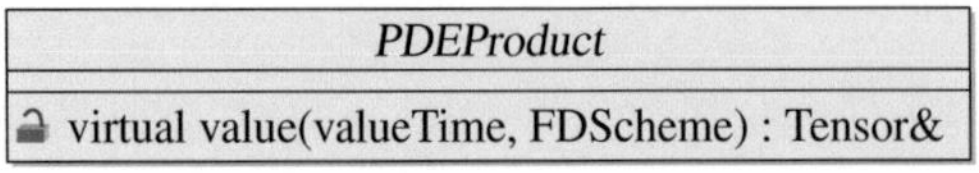

PDEProduct
🔒 virtual value(valueTime, FDScheme) : Tensor&

This method represents the product functionality and is implemented by every particular product type. This value-method has two arguments which are the value time at which the product is supposed to be valued and a reference on a FDScheme-Object. The method returns a reference on a tensor containing the calculated values.

In the following we discuss product routines for the product types discussed in Chapter 2.

Zero-Coupon Bond

A bond pays off one unit at its maturity and is straightforwardly valued by setting the final time values to be equivalent to one and roll back the scheme to value time.

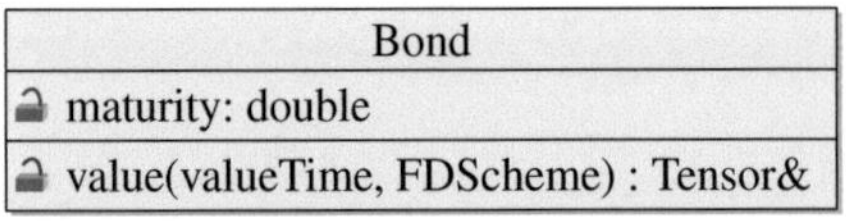

Bond
🔒 maturity: double
🔒 value(valueTime, FDScheme) : Tensor&

Figure 7.6.: The Product Class for the Zero-Coupon Bond

```
Tensor& value(double valueTime, FDScheme& scheme) {
  scheme.setFinalTime(maturity);
  scheme.values->apply(   Functionality(CONSTANT,1.0) );
  return scheme.rollback(valueTime);
}
```

Caplet

This class consists of the specific product properties such as the strike rate, its associated tenor the maturity of the caplet. Furthermore it im-

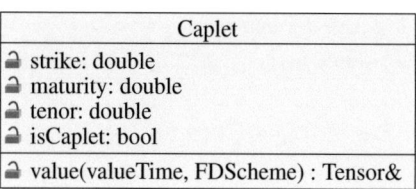

Caplet
🔒 strike: double
🔒 maturity: double
🔒 tenor: double
🔒 isCaplet: bool
🔒 value(valueTime, FDScheme) : Tensor&

Figure 7.7.: The Product Class for the Caplet

plements the method `value` with the particular valuation functionality: Firstly the final time of the corresponding boundary value problem is set to be the maturity of the caplet plus the tenor of the underlying term rate. The final time values are set to be one unit. This BVP is solved to the maturity of the caplet, where the bond-values get transformed into term-rates (according to formula 2.4). Afterwards we apply the Functor `Floor`, which corresponds to the option's payoff-profile. The resulting values become then rolled back to the desired value time.

```
Tensor& value(double ValueTime, FDScheme& scheme){
    //Start at Casflow-Time  T_i =  Maturity + Tenor
    scheme.setFinalTime(maturity+tenor);
    //Rollback to Maturity T_{i-1}
    scheme.values->apply( Functionality(CONSTANT,1.0) );
    scheme.rollback(maturity);
    //Local Copy for discountfactor
    Tensor& discountfactors = scheme.values->clone();
    //Convert to TermRate
    scheme.values->apply( Functionality(TERMRATE,tenor) );
    //Apply Payoff-Profile
    scheme.values->apply( Functionality(FLOOR,strike) );
    //Multiply with Tenor-Length and Discount to T_{i-1}
    scheme.getValues() *= discountfactors * tenor;
    discountfactors.~Tensor();
    scheme.setFinalTime(maturity);
    return scheme.rollback(valueTime);
}
```

Interest Rate Swap

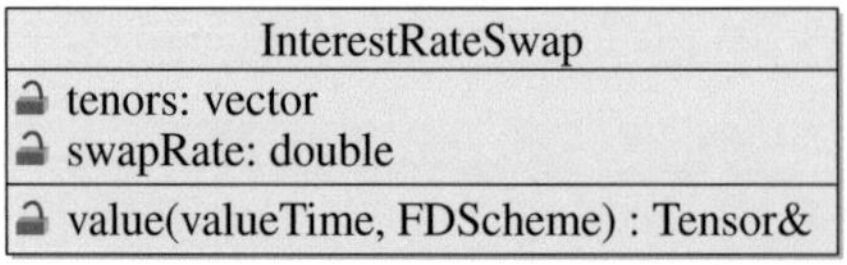

InterestRateSwap
▣ tenors: vector
▣ swapRate: double
▣ value(valueTime, FDScheme) : Tensor&

Figure 7.8.: The Product Class for the Interest Rate Swap

The valuation procedure for an interest rate swap is done for each tenor separately. The final time is set to the particular tenor point with final time profile constant 1. The scheme is rolled back to the previous tenor point - equal to the fixing time of the corresponding floating rate where the values get converted to term rates. After subtracting the swap rate the resulting values for this particular swap leg are rolled back to the first tenor point and are added on the temporary copy `swapValues`. After having repeated this for every tenor date the resulting values stored in `swapValues` are rolled back to the valuation time point.

```
Tensor& value(double valueTime, FDScheme& scheme){
  int len = tenors.size()-1;
  Tensor& swapValues = scheme.values->clone();
  swapValues.apply(Constant(0.0));
  for(int i= len ; i > 0 ; i-- ){
    //Start at Payment time point
    scheme.setFinalTime(tenors[i]);
    scheme.values->apply( Functionality(CONSTANT,1.0) );
    //Rollback to Fixing-Timepoint
    scheme.rollback(tenor[i-1]);
    //Local copy is needed for discounting
    Tensor& discountfactor = scheme.values->clone();
    //Calculate Term-Rate
    scheme.values->
        apply( Functionality(TERMRATE,tenor)   );
    scheme.getValues() -= swapRate; //Subtract SwapRate
    //Multiply with tenor and discount
    scheme.getValues() *=
              discountfactor * (tenor[i]-tenor1[i-1]) ;
    discountfactor.~Tensor();
```

```
    scheme.setFinalTime(tenors[i-1]);
    swapValues += scheme.rollback(tenors[0]);
  }
  scheme.setFinalTime(tenors[0]);
  scheme.setValues(swapValues);
  return scheme.rollback(valueTime);
}
```

European Swaption

The procedure to value a european (payer) swaption is straightforward. The underlying swap contract get valued at maturity and the Floor `functor` is applied on the corresponding values. The resulting values are rolled back the resulting values back to value time.

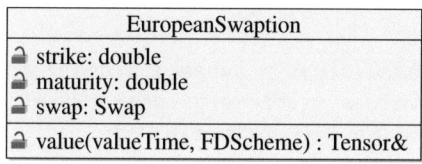

Figure 7.9.: The Product Class for the European Swaption

```
Tensor& value(double valueTime, FDScheme& scheme){
  swap.value(maturity,scheme).
                  apply( Functionality(FLOOR,strike) );
  scheme.setFinalTime(maturity);
  return scheme.rollback(valueTime);
}
```

Bermudan Swaption

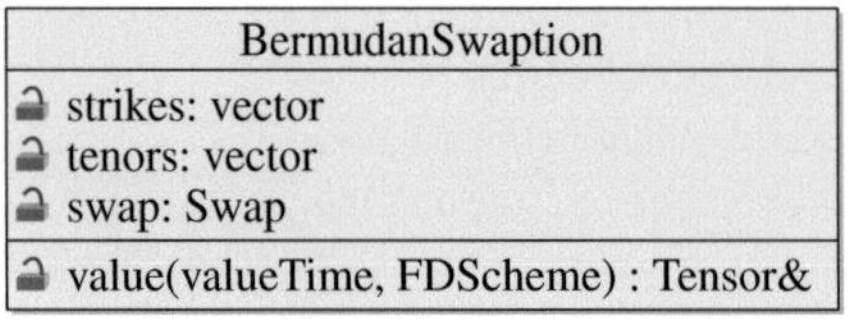

BermudanSwaption
⊒ strikes: vector
⊒ tenors: vector
⊒ swap: Swap
⊒ value(valueTime, FDScheme) : Tensor&

Figure 7.10.: The Product Class for the Bermudan Swaption

The procedure of a Bermudan Swaption works according to the valuation algorithm discussed in Sections 2.3 and 6.1. Note that we here apply a binary functor Max() applied to the swap values (Tensor swapValues) and bermudan values (Tensor bermudanValues) at each exercise date.

```
Tensor& value(double valueTime, FDScheme& scheme){
    int len = tenors.size()-1;
    //Local Copies
    Tensor& bermudanValues = scheme.values->clone();
    Tensor& swapValues = scheme.values->clone();
    bermudanValues.apply(Constant(0.0));
    swapValues.apply(Constant(0.0));
    for(int i = len; i>0 ; i--) {
        scheme.setValues(swapValues);       //Rollback Swap Leg
        scheme.setFinalTime(tenors[i]);
        scheme.rollback(tenors[i-1]);
        scheme.setValues(bermudanValues); //Rollback Bermudan
        scheme.setFinalTime(tenors[i]);
        scheme.rollback(tenors[i-1]);
        swapValues +=
            swap.getOnePeriod(tenors[i-1], tenors[i],scheme);
        bermudanValues =
            Tensor::apply(bermudanValues ,
                            Functionality(MAX) , swapValues);
    }
    scheme.setFinalTime(tenors[0]);
    scheme.setValues(bermudanValues);
    return scheme.rollback(valueTime);
}
```

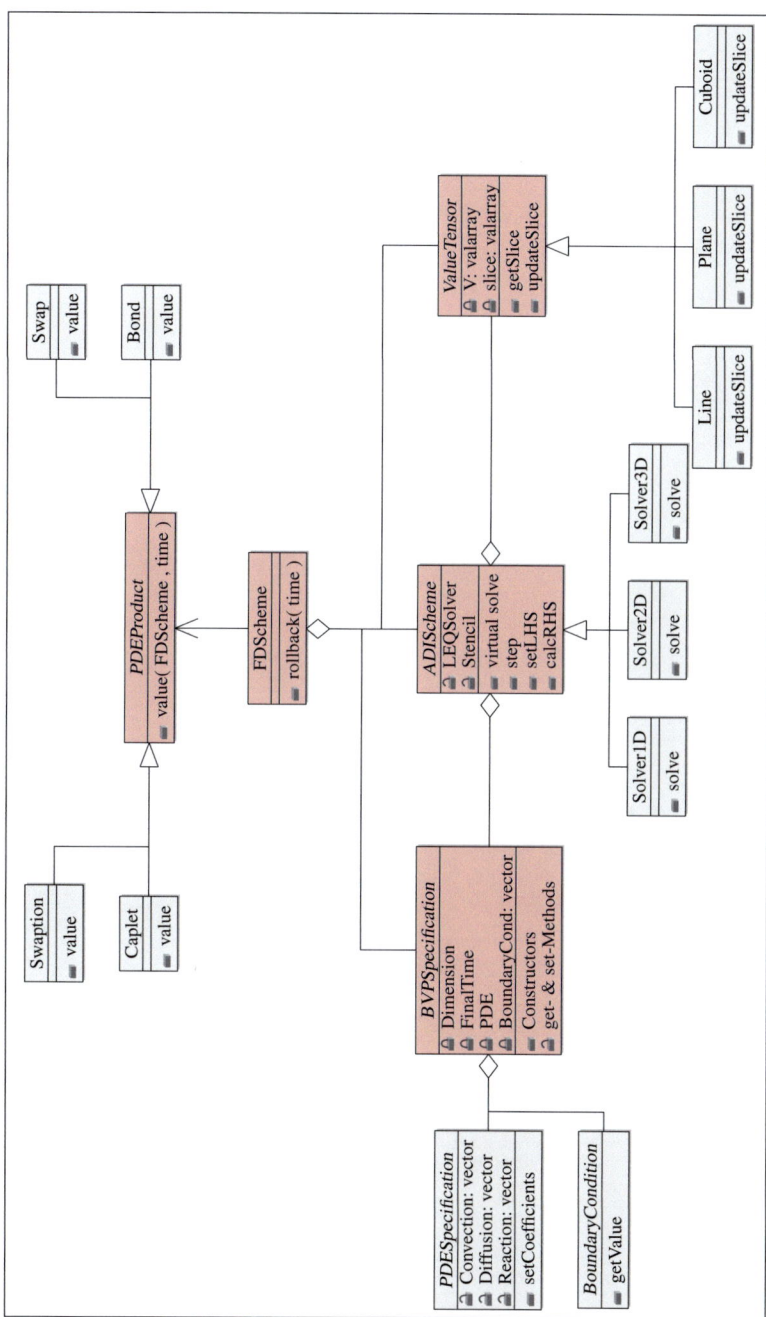

Figure 7.11.: C++ Design - Core Classes at a Glance

Conclusion

The main purpose of this book was to analyze a yield curve model with desirable numerical features to enable a deterministic PDE product valuation. In particular considering the robust numerical and implementation results we now reconsider the tree main objectives from the beginning:

⋄ In the theoretical treatment of the first chapters we reviewed the connection between specific stochastic process classes and partial differential equations in particular in view of financial valuation models. To properly embed the here analyzed yield curve model into the standard of interest rate modeling, we gave a detailed overview of the main product forms and model approaches.

⋄ The theoretical investigation of the exogenous Markovian yield curve model setup in Chapter 4 illuminated its close relation to the general HJM Framework and its increased flexibility compared to short rate models. Furthermore in Chapter 6 we gave an overview of how this flexibility can be effectively used in practice.

⋄ The discussed finite difference schemes from the fifth chapter enable us to efficiently cope with several versions of valuation PDE's (4.13, 4.16 and 6.5). Using alternative discretization methods we are able to overcome numerical difficulties due to missing diffusion parts or dominated convection. The generic approach of boundary handling is well suited to become applied to most product forms and even to other valuation models.

Reviewing the implementation side from the last chapter an object oriented design enabled us to detach the different functionalities - i.e numerical scheme, valuation model and product functionality.

In retrospective we have started with a formal definition of a stochastic process and ended up with the C++ implementation of the risk-neutral Bermudan Swaption valuation
- so far we let the matter rest here.

Part III.

Appendix

A. Additional Calculations

A.1. The Black-Scholes Model in a Nutshell

To give an important application of the concepts introduced in Section 1.8 we are now able to develop the whole Black Scholes context of valuing an european equity option under a constant interest-rate r.

⋄ On an arbitrary probability space, start with a lognormal type SDE of the form (1.21): $dS_t = \mu S dt + \sigma S dW^P$

⋄ It can be shown that an equivalent risk-neutral measure exists by applying Girsanov's Theorem (1.3) in combination with (1.40)

⋄ The resulting dynamics (1.39) imply that $d[D(s,t)S_t]$ is driftless therefore a martingale.

⋄ The conditional expectation

$$D(t,t)V_t = V_t = \mathbb{E}^{RN}\big[D(t,T)V(S_T^{RN},T)|\mathcal{F}_t\big]$$

is a martingale due to proposition (1.1)

⋄ The discounted time-t-value of an European Call Option with Strike K is then given by:

$$C(K,t) := \mathbb{E}^{RN}\big[D(t,T)(S_T^{RN} - K)_+|\mathcal{F}_t\big]$$

which can be calculated in closed form due to the analytical solution of SDE (1.21). The formula can be found e.g. in [8, p.91].

⋄ Due to the markovian structure of the used SDE we can apply Feynman-Kac Theorem (1.4) which yields the well-known parabolic *Black-Scholes PDE*:

$$\frac{\partial V}{\partial t} + \frac{\partial V}{\partial S}rS + \frac{\partial^2 V}{\partial S^2}\sigma^2 S^2 - rV = 0 \qquad (A.1)$$

with final-time condition: $V(S,T) := (S - K)_+$

A.2. Evaluating Expected Values and Black's Formula

The following Lemma is very useful, when it comes to evaluate expected values such as the risk-neutral expectation formula, such as the Black-Scholes or Black's Formula.

Lemma A.1. *Let* $X \sim \mathcal{N}(\mu, \sigma^2)$*. Then for any* $a, K \in \mathbb{R}$

$$\mathbb{E}\big[e^{aX} \text{ with } X \geq K\big] = e^{a\mu + \frac{1}{2}a^2\sigma^2} N(d) \tag{A.2}$$

where

$$d := \frac{-K + \mu + a\sigma^2}{\sigma}$$

Remark A.1. *We know, that if* X *has gaussian distribution, then* e^X *is lognormal distributed. As we now the mean of* e^X *explicitly we give a second version of the above formula:*

$$\mathbb{E}\big[e^X \text{ with } X \geq K\big] = \mathbb{E}[e^X]N(d)$$

where

$$d := \frac{\log\left(\frac{e^{\mu + \frac{1}{2}\sigma^2}}{e^K}\right) + \frac{1}{2}\sigma^2}{\sigma} = \frac{\log\left(\frac{\mathbb{E}[e^X]}{e^K}\right) + \frac{1}{2}\sigma^2}{\sigma}$$

Now we can easily state *Black's Formula* which we introduced in a more abstract form in Chapter 2.

Theorem A.1 (Black's Formula). *Let* X *be lognormally distributed with Mean* $F := \mathbb{E}^T[X]$ *and Variance* σ^2*. Then under the Forward Measure:*

$$\mathbb{E}^T\big[[X - K]_+\big] = \mathbb{E}^T[X]N(d_1) - KN(d_2) \tag{A.3}$$

with

$$d_1 := \frac{\log\left(\frac{F}{K}\right) + \frac{1}{2}\sigma^2}{\sigma} \quad \text{and} \quad d_2 := d_1 - \sigma$$

A.3. Expressing a Caplet in Zero-Coupon Bond Puts

With the following proposition we express a caplet as a portfolio of zero-bond options.

Proposition A.1. *A Caplet with time interval $[T_1, T_2]$ and strike rate K corresponds to a multiple of zero-coupon bond options with strike price P_K:*

$$\mathrm{Cpl}_{T_1, T_2} = (1 + K\Delta_2)(P_K - P(T_1, T_2))_+$$

Proof. Start with the caplet payoff at time T_2 with a future term rate $L(T_1, T_2)$:

$$\mathrm{Cpl} = \Delta_2 \max(L(T_1, T_2 - K, 0)$$

which we can discount to time T_1 with discount factor $(1 + L(T_1, T_2)\Delta_2)$

$$
\begin{aligned}
&= \max\left(\frac{\Delta_2(L(T_1, T_2) - K)}{1 + \Delta_2 L(T_1, T_2)}, 0\right)\\
&= \max\left(\frac{1 + \Delta_2 L(T_1, T_2)}{1 + \Delta_2 L(T_1, T_2)} - \frac{1 + \Delta_2 K}{1 + \Delta_2 L(T_1, T_2)}, 0\right)\\
&= \max\left(1 - \frac{1 + \Delta_2 K}{1 + \Delta_2 L(T_1, T_2)}, 0\right)\\
&= \max\left(1 - (1 + \Delta_2 K) P(T_1, T_2), 0\right)
\end{aligned}
$$

Factoring out $(1 + \Delta_2 K)$ and using the relationship between bonds and term rates (2.4) yields

$$= (1 + \Delta_2 K) \max\left(\underbrace{\frac{1}{1 + \Delta_2 K}}_{:=P_K} - \underbrace{\frac{1}{1 + \Delta_2 L(T_1, T_2)}}_{=P(T_1, T_2)}, 0\right)$$

$\square$

Combining this with the analytical formula for bond options in the specific context of Section 4.4, we will therefore also have an analytical caplet value.

A.4. Expressing a Swaption as a Sum of Put Options

We already noted that since the swaption payoff (2.24) is a convex function, it can not additively decomposed into its singular components. In the short-rate context (3.2) there is a difference:

$$P(t,T) = P(t,T;r_t) := A(t,T)\exp[-B(t,T)r_t]$$

Put it another way: bonds with different maturities are *monotone decreasing functions* of the current short-rate r_t at time t. We again look at the payoff of a payer swaption at exercise time T_E which is here equal to the first tenor date of the underlying swap $T_E = T_0$:

$$\Phi_{\text{Swaption}} = \left[1 - P(T_0, T_N; r_t) - \sum_{i=1}^{N} P(T_0, T_i; r_t)\Delta_i R\right]_+$$

As we know from above that every bond value depends on $r(t)$ only, each single summand of entire payoff function becomes a function depending on $r(t)$. This also implies that there exists a value of the short rate $\bar{r}$ with

$$\sum_{i=1}^{N} P(T_0, T_i; \bar{r})\Delta_i R = 1$$

For any $r_t > \bar{r}$ each single summand will be positive. Therefore we can decompose the above payoff which yields:

$$\Phi_{\text{Swaption}} = \left[P(T_0, T_N; \bar{r}) - P(t, T_N; r_t)\right]_+$$
$$+ R\sum_{i=1}^{N} \Delta_i\left[P(T_0, T_i; \bar{r}) - P(T_0, T_i; r_t)\right]_+$$

So the payoff of a swaption is nothing else than the sum of put options on zero-coupon bonds.

A.5. Product Specific Valuation PDEs

Although we developed generic pricing routines for several product forms in Chapter 7, we would like to emphasize that only with minor changes in the reaction terms of a valuation PDE we are able to value specific product forms directly. To unify that procedure we assemble all versions the valuation PDE's (4.16, 4.13 and 6.5) by writing:

$$\frac{\partial V}{\partial t} + \mathcal{L}^{RN} V - rV = 0$$

Here $\mathcal{L}^{RN}$ denotes the specific spatial differential operator under the risk-neutral dynamics. We now give the product specific valuation PDE's of an interest rate swap and an interest rate cap:

Proposition A.2. *The risk-neutral conditional expectation V_t of an interest rate swap with swap rate R satisfies:*

$$\frac{\partial V}{\partial t} + \mathcal{L}^{RN} V - rV + (r - R) = 0$$

with final-time profile $\Phi(x) \equiv 0$.
For an interest rate cap with strike rate K we have:

$$\frac{\partial V}{\partial t} + \mathcal{L}^{RN} V - rV + \min[r - K, 0] = 0$$

with final-time profile $\Phi(x) \equiv 1$.

These specific PDEs are consistent with the according valuation routines from Chapter 7. Note that we here have to modify the valuation PDEs, which is from our view not preferable in the generic implementation setting from Chapter 7.

A.6. Finite Differences: Further Discretizations

Backward Discretizations

The here described second-order accurate discretization involves the grid points x_{m-2}, x_{m-1}, x_m: Consider the following Taylor expansions:

$$U_{m-1} = U_m - \frac{\partial u}{\partial x}h + \frac{1}{2}\frac{\partial^2 u}{\partial x^2}h^2 + O(h^3)$$

$$U_{m-2} = U_m - 2\frac{\partial u}{\partial x}h + 2\frac{\partial^2 u}{\partial x^2}h^2 + O(h^3)$$

Solving for the first derivative yields:

$$\frac{\partial u}{\partial x} = \frac{U_{m-2} - 4U_{m-1} + 3U_m}{2h} + O(h^2)$$

This is known as a second order *backward* discretization.

Five Point Discretizations

As an example of a discretization involving more than three grid points we can obtain fourth order accuracy when considering additional and higher Taylor Expansions for the additional points x_{m-2} and x_{m+2} (we write subscripts instead of partial derivatives here):

$$U_{m+2} - U_m = 2\partial_x uh + 2\partial_{xx}uh^2 + \frac{4}{3}\partial_{xxx}uh^2 + \frac{4}{5}\partial_{xxxx}uh^4 + O(h^5)$$

$$U_{m-2} - U_m = -2\partial_x uh + 2\partial_{xx}uh^2 - \frac{4}{3}\partial_{xxx}uh^3 + \frac{4}{5}\partial_{xxxx}uh^4 + O(h^5)$$

$$U_{m+1} - U_m = \partial_x uh + \partial_{xx}uh^2 + \frac{1}{6}\partial_{xxx}uh^3 + \frac{1}{24}\partial_{xxxx}uh^4 + O(h^5)$$

$$U_{m-1} - U_m = -\partial_x uh + \partial_{xx}uh^2 + \frac{1}{6}\partial_{xxx}uh^3 + \frac{1}{24}\partial_{xxxx}uh^4 + O(h^5)$$

Solving - i.e. canceling higher derivatives - yields for the first derivative

$$\frac{\partial u}{\partial x} = \frac{U_{m-2} - 8U_{m-1} + 8U_{m+1} - U_{m+2}}{12h} + O(h^4)$$

B. Probability Essentials

To have most notions well-defined we have collected here used concepts and definitions from probability theory. For that purpose we made use of the books [6] and [4].

Probability Spaces

⋄ A *Probability Space* $(\Omega, \mathcal{F}, P)$ is equipped with a *measurable space* $(\Omega, \mathcal{F})$ consisting of a chosen set Ω with elements ω and a σ-Algebra $\mathcal{F}$ defined on Ω as the set of all admissible subsets and a probability measure P.

⋄ A Probability Measure is a function $P : \mathcal{F} \mapsto [0, 1]$ satisfying the according probability axioms:

  ⋄ $P(\emptyset) = 0$ and $P(\Omega) = 1$

  ⋄ Countable Additivity: $P(\cup_i A_i) = \sum_i P(A_i)$ for $A_i \in \mathcal{F}$ pairwise disjoint

⋄ If $\Omega := \mathbb{R}^n$ then the according σ-Algebra is referred to as Borel-σ-Algebra as the set of all open subsets $B \subset \mathbb{R}^n$ and is denoted with $\mathcal{B}(\mathbb{R}^n)$.

Random Variables

⋄ Given a probability space $(\Omega, \mathcal{F}, P)$ a function $X : \Omega \mapsto \mathbb{R}^n$ is called $\mathcal{F}$-measurable if for all open subsets $U \subset \mathbb{R}^n$

$$X^{-1}(U) := \{\omega \in \Omega | X(\omega) \in U\} \in \mathcal{F} \tag{B.1}$$

⋄ A n-dimensional random variable or *random vector* $X = (X_1, \ldots, X_n)$ is a $\mathcal{F}$-measurable function from $(\Omega, \mathcal{F}, P) \mapsto (\mathbb{R}^n, \mathcal{B}(\mathbb{R})^n)$
A random variable is characterized by events of the form: $\{\omega \in \Omega | X(\omega) \in B\} \in \mathcal{F}$.

⋄ The σ-algebra generated by a random variable X is given by:

$$\sigma(X) = \{A \subset \Omega | X^{-1}(B) = A \text{ for some } B \in \mathcal{B}\} \qquad \text{(B.2)}$$

It can be shown (see [6, p.195]) that Y is measurable w.r.t. to $\sigma(X)$ if and only if there exists a measurable function f on $\mathbb{R}^n$ such that $Y = f(X)$. In example X is $\sigma(X)$-adapted.

⋄ In the context of stochastic processes Y_t is $\sigma(X)$-measurable, if there exists a Borel measurable function $f : \mathbb{R} \to \mathbb{R}$ such that $Y = f(X)$

⋄ On $(\Omega, \mathcal{F}, P)$ a random variable $X : \Omega \mapsto \mathbb{R}$ introduces a probability measure P^X on $(\mathbb{R}, \mathcal{B})$. P^X is called the *Distribution* of X. For any $B \in \mathcal{B}$:

$$P^X(B) = P(X \in B) = P(X^{-1}(B)) = P(\{\omega \in \Omega : X(\omega) \in B\}). \qquad \text{(B.3)}$$

⋄ *Independence*: Two random variables X and Y are to be said independent, if:

$$P(X^{-1}(A) \cap Y^{-1}(B)) = P(X^{-1}(A))P(Y^{-1}(B)) \quad \forall A, B \in \mathcal{B} \qquad \text{(B.4)}$$

⋄ The function $F : \mathbb{R} \mapsto [0,1]$:

$$F^X(x) := P^X\big((-\infty, x]\big)$$

is called the *Distribution Function* of X.

⋄ Extension: The *joint distribution function* $F : \mathbb{R}^n \mapsto [0,1]$ is given by:

$$F^X_{x_1,\dots,x_n} = P(X_1 \leq x_1, \dots X_n \leq x_n) = P\big(\prod_{i=1}^{n}(-\infty, x_i]\big) \quad \text{(B.5)}$$

⋄ A *Probability Density* is a Lebesque integrable function $f : \mathbb{R}^n \mapsto [0,\infty)$ such that

$$F_{x_1,\dots,x_n} := \int_{-\infty}^{x_1} \dots \int_{-\infty}^{x_n} f(x_1, \dots, x_n) dx_1 \dots dx_n \qquad \text{(B.6)}$$

Moments of Random Variables

◇ First Moment: *Expectation* of a random variable X w.r.t. the probability measure P:

$$\mathbb{E}^P[X] = \int_\Omega X(\omega)dP(\omega) = \int_\mathbb{R} x\,dP^X(x) \qquad (B.7)$$

◇ If we have a random variable that maps into a countable space $X : \Omega \mapsto \mathcal{S}$ we have:

$$\mathbb{E}^P[X] = \sum_{x \in \mathcal{S}} xP(X = x)$$

with $P(X = x) = P\big(\{\omega \in \Omega \big| X(\omega) = x\}\big)$

◇ For a Borel measurable and integrable function f we have:

$$\mathbb{E}^P[f(X) \in A] = \int_A f(x)dP^X(x) = \int_{X^{-1}(A)} f(X(\omega))dP(\omega)$$

◇ The function class of functions for which $\mathbb{E}^P[a(X)] < \infty$ is denoted as $\mathcal{L}_1$

◇ If X and Y are independent $\Rightarrow \mathbb{E}[X_1 X_2] = \mathbb{E}[X_1]\mathbb{E}[X_2]$

◇ Relation between Measure and Expectation: Choosing $f(.) := \mathbf{1}(.)$ yields:

$$P^X(A) = \mathbb{E}^P[\mathbf{1}_A] = \int_\mathbb{R} \mathbf{1}_A(x)dP^X(x) = \int_\Omega \mathbf{1}_{X^{-1}(A)}(\omega)dP(\omega) \qquad (B.8)$$

◇ Second Moments: *Variance, Covariance and Correlation*

$$\text{Var}(X) := \mathbb{E}\big[X - \mathbb{E}[X]^2\big], \quad \sqrt{\text{Var}(X)} := \sigma(X)$$
$$\text{Cov}(X,Y) := \mathbb{E}\big[X - \mathbb{E}[X]\big]\mathbb{E}\big[Y - \mathbb{E}[Y]\big] \quad X,Y \text{ Random Variables}$$

For $X = [X_1, \ldots, X_n] \in \mathbb{R}^n$ we have the positive-semidefinite *Covariance-Matrix*:

$$C \in \mathbb{R}^{n \times n}, \quad [C]_{i,j} := \text{Cov}(X_i, X_j) \quad \forall i, j = 1, \ldots, n$$

and *Correlation-Matrix* ρ with elements:

$$\rho_{ij} := \frac{Cov(X_i, X_j)}{\sigma X_i \sigma X_j}, \quad -1 \leq \rho_{ij} \leq 1$$

◇ If X_i and X_j are independent: $\Rightarrow Cov(X_i, X_j) = 0 \Rightarrow \rho_{ij} = 0$

◇ If X_i are pairwise uncorrelated we have $Var(\sum_i X_i) = \sum_i X_i$

◇ If X is a n-dim random variable with Covariance Matrix C_X, then $Y := AX$, $A \in \mathbb{R}^{n \times n}$ is a random variable with Covariance $C_Y := ACA^t$

Gaussian Random Variables

◇ A 1-dimensional random variable X is said to be normally distributed: $X \sim \mathcal{N}(\mu, \sigma)$, if its density is defined by:

$$f(x, \mu, \sigma) := \frac{1}{\sqrt{2\pi}\sigma} \exp\left(\frac{-(x-\mu)^2}{2\sigma^2}\right) \tag{B.9}$$

◇ Cumulative Distribution Function:

$$N(x|\mu, \sigma) = \int_{-\infty}^{x} \frac{1}{\sqrt{2\pi}\sigma} \exp\left(\frac{-(x-\mu)^2}{2\sigma^2}\right) dx \tag{B.10}$$

◇ $\sum_{i=1}^{n} X_i \sim \mathcal{N}(\sum_{i=1}^{n} \mu_i, \sum_{i=1}^{n} \sigma_i^2)$ for X_i pairwise independent

◇ $X \sim \mathcal{N}(\mu, \sigma^2) \Rightarrow aX + b \sim \mathcal{N}(a\mu + b, a^2\sigma^2)$

◇ *Central Limit Theorem* (very roughly)

$$\sum_{i=1}^{n} \frac{X_i - \mu}{\sqrt{n}\sigma} \to \mathcal{N}(0, 1) \tag{B.11}$$

for X_i independent identically distributed random variables with mean μ and Variance σ^2.

◇ A n-dimensional random variable X is said to be *Multivariate Normal* $X \sim \mathcal{N}(\mu, C)$, if $a^T X \in \mathbb{R}$ is normally distributed $\forall a \in \mathbb{R}^n$ with *mean-vektor* μ and Covariance C

Stochastic Convergence

⋄ **Convergence in Probability:** A Sequence of RV $(X_n)_{n\geq 1}$ converges to X in probability if:

$$\lim_{n\to\infty} P\big(\{\omega : |X_n(\omega) - X(\omega)| > \epsilon\}\big) = 0 \qquad \text{(B.12)}$$

⋄ **Convergence in Distribution:** A Sequence of RV $(X_n)_{n\geq 1}$ converges weakly to X $(X_n \to_d X)$ if for all continuous bounded functions f:

$$\lim_{n\to\infty} \int_{\mathbb{R}} f(x) dP^{X_n}(x) = \int_{\mathbb{R}} f(x) dP^X(x)$$

⋄ **Mean-Square Convergence:** A Sequence of RV $(X_n)_{n\geq 1}$ converges to X in the mean-square limit if:

$$\lim_{n\to\infty} \mathbb{E}\big[|X_n - X|^2\big] = 0 \qquad \text{(B.13)}$$

Conditional Expectation

Given a probability space $(\Omega, \mathcal{F}, P)$ and $\mathcal{G} \subset \mathcal{F}$.

Definition B.1. *The* Conditional Expectation *of a random variable* $X : \Omega \to \mathbb{R}^n$ *given* $\mathcal{G} \subset \mathcal{F}$ *is itself a random variable:*

$$\mathbb{E}[X|\mathcal{G}] : \Omega \to \mathbb{R}^n \qquad \text{(B.14)}$$

where it satisfies:

1. $\mathbb{E}[X|\mathcal{G}]$ *is* $\mathcal{G}$ *measurable*

2.

$$\int_G \mathbb{E}[X|\mathcal{G}] dP = \int_G X dP \quad \forall \quad G \in \mathcal{G}$$

We give some further basic properties of (B.14), where X, Y denote random variables.

$\diamond$ $X \mapsto \mathbb{E}[X|\mathcal{G}]$ is linear

$\diamond$ $\mathbb{E}[\mathbb{E}[X|\mathcal{G}]] = \mathbb{E}[X]$

$\diamond$ $\mathbb{E}[X|\mathcal{G}] = X$ if X is $\mathcal{G}$ measurable

$\diamond$ $\mathbb{E}[X|\mathcal{G}] = E[X]$ if X is independent of $\mathcal{G}$

$\diamond$ $\mathbb{E}[YX|\mathcal{G}] = YE[X|\mathcal{G}]$ if Y is $\mathcal{G}$ measurable

Proposition B.1 (Tower Property). *Let there be σ-Algebras $\mathcal{H}$ and $\mathcal{G}$ such that $\mathcal{H} \subset \mathcal{G} \subset \mathcal{F}$. Then:*

$$\mathbb{E}[\mathbb{E}[X|\mathcal{H}]|\mathcal{G}] = \mathbb{E}[X|\mathcal{G}] \tag{B.15}$$

Bibliography

[1] O. Cheyette. Markov Representation of the Heath-Jarrow-Morton Model. Available at SSRN: http://ssrn.com/abstract=6073, August 1996.

[2] J. Andreasen. *Back To The Future*. *RISK*, pages 104–109, September 2005.

[3] J. Steele. *Stochastic Calculus and Financial Applications*. Springer-Verlag, New York, 2001.

[4] B. Oksendal. *Stochastic Differential Equations*. Springer-Verlag, sixth edition, 2003.

[5] J. Strikwerda. *Finite Difference Schemes and Partial Differential Equations*. 1989.

[6] J. Jacod and J. Protter. *Probability Essentials*. Springer-Verlag, 2000.

[7] S. Karatzas and S. Shreve. *Brownian Motion and Stochastic Calculus*. Springer-Verlag, New York, 1988.

[8] M. Baxter and A. Rennie. *Financial Calculus - An Introduction to Derivative Pricing*. Cambrigde University Press, 1996.

[9] D. Brigo and F. Mercurio. *Interest Rate Models - Theory and Practice*. Springer Verlag, Heidelberg, 2001.

[10] M. Musiela and M. Rutkowski. *Martingale Methods in Financial Modelling*. Springer Verlag, 2004.

[11] C. Fries. *Mathematical Finance: Theory, Modeling, Implementation*. Wiley, 2007.

[12] R. Rebonato. *Term Structure Models: A Review*. Royal Bank of Scotland Quantitative Research Centre Working Paper, 2003.

[13] M. Avellaneda and P. Laurence. *Quantitative Modeling of Derivative Securities.* Chapman Hall, 2000.

[14] J. Hull. *Options, Futures and other Derivatives.* Prentice Hall, fifth edition, 2003.

[15] L. Andersen and R. Brotherton-Ratcliffe. *Extended Libor Market Models with Stochastic Volatility. Mathematical Finance,* 5(1):55–72, 1995.

[16] P. Ritchken and L. Sanakarasubramanian. *Volatility Structures of Forward Rates and the Dynamics of the Term Structure. Mathematical Finance,* 5(1):55–72, 1995.

[17] J. Andreasen. *Turbo Charging the Cheyette Model.* Working Paper GenRe Securities, 2000.

[18] I. Craig and S. Sneyd. *An Altering Implicit Scheme for Parabolic Equations with Mixed Derivatives. Computers and Mathematics with Applications,* 16(4):341–350, February 1988.

[19] K. Morton and D. Mayers. *Numerical Solution of Partial Differential Equations.* Cambridge University Press, 1994.

[20] J. Thomas. *Numerical Partial Differential Equations.* Springer, 1995.

[21] H. Kushner and P. Dupuis. *Numerical Methods for Stochastic Control Problems in Continuous Time.* Springer, 2001.

[22] A. Mitchell and D. Griffiths. *The Finite Difference Method in Partial Differential Equations.* John Wiley and Sons, 1980.

[23] T. Domingo and C. Randall. *Pricing Financial Instruments: The Finite Difference Method.* Wiley, 2000.

[24] J. Gatheral. *The Volatility Surface.* Whiley, 2006.

[25] M. Meister. Smile Modeling in the Libor Market Model. Diploma thesis, University of Karlsruhe, August 2004.

[26] J. Andreasen and L. Andersen. Volatile Volatilities. *RISK,* pages 163–168, December 2002.

Index

Alternating Direction Implicit Scheme (ADI)
Application to Valuation PDE, 129
C++ Implementation, 169
Craig & Sneyd Scheme, 126
General Idea, 123
Local Accuracy and Stability, 131
Peaceman-Rachford Scheme, 124
Arbitrage-Free Model, 31

Bellman's Backward Induction, 151
BGM-Model, *see* Lbor Market Model 70
Black's Formula, iv
Black-Scholes Model, iii
Boundary Conditions
Appropriate Space Domains, 141
Dirichlet, 137
Linear Extrapolation, 138
Error introduced by, 139
Neumann, 137
Numerical, 138
Robbins, 137
Truncation Error, 141
Brownian Motion
Definition, 11
Forward, 52
Geometric, 20
Girsanov Transformation, 23
Heat Equation, 12
Multiple Dimensions, 13
Transition Density, 11

Cheyette Model
Constant Parameter Case, 91
One-Factor Version, 79
Achieving Consistency with LMM, 156
Analytical Formulae, 93
Bond Price Formula, 85
C++ Implementation, 176
Derivation from HJM, 80
Markovian Short-Rate, 83
Markovian Yield-Curve, 77
Motivation, 77
Multi-Factor Version, 95
Numerical Treatment, 134
Parameter Interpretation, 84
Valuation PDE, 89, 157
Numerical Treatment, 99
with stochastic volatility, 157

Diffusion Process
Transition Density, 10
Forward PDE, 10

Early Excercise Problem, 151

Feynman-Kac PDE
  Finite Differences, 113
  Relation to SDE's, 104
Feynman-Kac Theorem, 26
  Application to Risk-Neutral
    Valuation, 89
  Numerical Treatment, 99
  Proof of, 27
  Relation to Backward Equation, 8
  Relation to Conditional Expecation, 26
Finite Difference Scheme
  Application to Feynman-Kac
    PDE, 103
  C++ Implementation, 168
  Consistency, 115
  Convergence, 114
  Explicit Scheme, 106
  Five-Point Discretization, viii
  General Procedure, 105
  Grid Transformation, 146
  Implicit Scheme, 107
  Incoporation of Boundary Conditions, 136
  Lax-Richtmeyer Equivalence
    Theorem, 114
  Lax-Wendroff Scheme, 110
  Non-Uniform Discretization, 145
  Numerical Boundary Condition, 136
  Relation to Markov Chains, 113
  Stability, 118
  Thomas Algorithm, 101
  Three-Point Discretizations, viii, 100
  Unsymmetric Differences, 145

  Upwind Scheme, 108
  Weighted Scheme, 107
Fixed-Income Market
  Bermudan Swaption, 49
    as an early-exercise problem, 151
    C++ Implementation, 184
  Black's Formula, iv, 56
  Bond Option, 45
  Caplet, 46
    C++ Implementation, 181
    Expression in Bond Options, v
    Valuation PDE, vii
  Derivatives
    Digital Caplet, 46
    Early-Exercise Products, 48
  European Swaption, 47
    C++ Implementation, 183
    Expression in Bond Options, vi
  Exotic Derivatives, 48
  Floating Rate Notes, 42
  Forward Measure, 52
    Valuation under, 55
  Forward Rate
    Construction, 39
    Relation to Short Rate, 54
  Interest Rate Swap, 43
    C++ Implementation, 182
    Valuation PDE, vii
  Market versus Model, 55
  Product Specific Valuation
    PDEs, vii
  Risk-Neutral Valuation, 50
  Short Rate, 38
  Tenor Structure, 42

Term Rate, 38
Valuation in Black's Model,
    56
Yield, 38
Yield Curve
    Definition, 40
    Risk-Neutral Version, 53
Zero-Coupon Bond, 38
    C++ Implementation, 180
Forward-Rate Volatility
    Effect of Mean-Reversion, 84
    in Hull-White Model, 68
    in Libor Market Model, 72
    Local Volatility Approach,
        153
    Multifactor, 95
    Separable Structure, 79
    Stochastic Volatility, 156

Girsanov's Theorem, 23

Heath-Jarrow Morton (HJM)
    Discussion, 68
    Forward-Rate Dynamics, 64
    Implied Short-Rate Dynam-
        ics, 67
    Log-Normal Explosion, 69
    Motivation, 63
    Multi-Factor Version, 68
    PDE Valuation, 89
    Relation to Libor Market Model,
        68
    Relation to Short-Rate Mod-
        els, 68
    Separable Volatility Struc-
        ture, 79
Hull-White Model
    as an instance of the Cheyette
        Model, 91

as instance of HJM Model,
    68
Derivation, 91
Dynamics, 60
Valuation PDE, 94

Implementation
    ADI Scheme, 169
    Back End, Front End, 167
    Class for Difference Opera-
        tors, 173
    Functor Concept, 175
    Grid Class, 171
    handling of Boundary Con-
        ditions, 172
    handling of several dimen-
        sions, 174
    PDE Product Class, 180
    Tensor class, 174
    UML Diagram, 186
Implied Volatility
    Concept, 162
    implied by model, 162
    Smile, 162
Ito Calculus
    Ito Integral
        Construction of, 14
        Definition, 15
    Ito Isometry, 16
    Ito Process, 16
    Ito's Lemma, 17
    Motivation, 14
    Product Rule, 18

Libor Market Model
    Derivation from HJM Dy-
        namics, 71
    Forward-Rate Dynamics, 74
    Measure-Change, 74
    Motivation, 70

PDE Valuation, 75
Spot-Libor Measure, 74
Local Volatility, 153
CEV Specifiation, 155
Displaced Diffusion Approach, 155
Parametrization, 154

Market Completeness, 33
Markov Process
Backward Equation, 8
Definition, 7
Dynamics of Conditional Expectation, 8
Forward Equation, 8
Markov Chain, 7
Relation to Finite Difference Schemes, 113
Transition Probability, 7
Markovian Yield-Curve
PDE Valuation, 89
Martingale
Construction, 5
Definition, 5
Mean-Reversion, 22
Effect on volatility, 84

Numeraire
Annuity Numeraire, 57
Bond Numeraire, 52
Change-of-Measures, 35
Change-of-Numeraire Technique, 34
Definition, 34
Money-Market, 32
Rolling Deposit, 70
Spot Libor, 70

Optimal Control Problems, 151
Optimality Principle, 151

Partial Differential Equation
Convection-Diffusion Behavior, 104
Dynamics of Conditional Expectation, 26
Feynman-Kac PDE, 26
Heat Equation, 12
Risk-Neutral Valuation, 89
Short-Rate Models, 62
Transition Densities, 10
Probabilitiy Measure
Construction of, 25
Equivalent, 23
Forward Risk-Neutral, 52
Risk-Neutral, 32
Spot-Libor, 74
Transformation of, 23

Radon-Nikodym Theorem, 23
Risk-Neutral Valuation, 32

SDE
Relation to Stochastic Processes, 20
Definition, 19
Lognormal, 20
Mean-Reversion, 22
Ornstein-Uhlenbeck Process, 20
Short-Rate Models
as instances of HJM Framework, 68
Bond Valuation, 60
Common Properties, 60
Drawbacks, 61
General Procedure, 62
PDE Valuation, 62
Risk-Neutral Yield Curve, 61

Stochastic Differential Equation
(SDE), 19
Stochastic Process
σ-Algebra generated by, 4
Adapted Process, 5
Filtration, 4
Information Structure, *see*
Filtration
Path-Independent, 19
State Space, 3
Stochastic Volatility Model, 156
Process Dynamics, 156
true stochastic volatility, 159

Trees
Binomial Tree, 9
Decision Tree, 9
Path Independence, 9
Relation to Stochastic Pro-
cesses, 9

Vasicek Model
Dynamics, 60
von-Neumann Stability Analysis, 118